The Age of Diagnosis

About the Author

Dr Suzanne O'Sullivan has been a consultant in neurology since 2004, first working at The Royal London Hospital and now as a consultant in clinical neurophysiology and neurology at The National Hospital for Neurology and Neurosurgery, and for a specialist unit based at the Epilepsy Society. She specialises in the investigation of complex epilepsy and also has an active interest in psychogenic disorders. Suzanne's first book, *It's All in Your Head*, won both the Wellcome Book Prize and the Royal Society of Biology Book Prize and *The Sleeping Beauties* was shortlisted for the Royal Society Science Book Prize. She is from Dublin, Ireland.

SUZANNE O'SULLIVAN

The Age of Diagnosis

Sickness, Health and Why Medicine Has Gone Too Far

hodder
press

First published in Great Britain in 2025 by Hodder Press
An imprint of Hodder & Stoughton Limited
An Hachette UK company

The authorised representative in the EEA is Hachette Ireland, 8 Castlecourt Centre,
Dublin 15, D15 XTP3, Ireland (email: info@hbgi.ie)

A CIP catalogue record for this title is available from the British Library

Hardback ISBN 9781399727648
Trade Paperback ISBN 9781399727655
ebook ISBN 9781399727679

Typeset in Plantin MT Pro by Manipal Technologies Limited

Printed and bound in Great Britain by Clays Ltd, Elcograf S.p.A.

Hodder & Stoughton policy is to use papers that are natural, renewable and recyclable products and made from wood grown in sustainable forests. The logging and manufacturing processes are expected to conform to the environmental regulations of the country of origin.

Hodder & Stoughton Limited
Carmelite House
50 Victoria Embankment
London EC4Y 0DZ

www.hodderpress.co.uk

Contents

Author's Note

All the stories in this book are real. I recount them to you much as they were told to me, edited for length but not for story. However, because of the deeply personal nature of some of what was discussed, many of the names and identifying details have been changed. Some are patients under my care and when that is the case, I make it clear in the text. I have also spoken to a range of experts for each topic, some of whom asked not to be named because of the sensitive nature of the discussions.

Prologue

I have thought of Abigail often over the past decade, always with a slight feeling of guilt. Because when she was only fifteen, long before I had even met her, I did something that might have changed her life forever.

'Did you believe that you had something medically wrong with you when I suggested you see a neurologist?' I asked her recently.

'Not really,' she laughed.

That had been my worry. That I had unnecessarily medicalised the life of a teenage girl.

Abigail is now a nursery worker in her mid-twenties. Back in 2012, without giving it nearly enough thought, I had given her a diagnosis she hadn't asked for. This is what I'm trained to do as a doctor. It is the focus of medicine right now: to anticipate future diagnoses, to screen for disease, to diagnose at ever earlier stages. But, looking back all these years later, I worry that I turned her into a patient for no good reason.

Abigail's mother, Stephanie, has been my patient for nearly twenty years. She has faced every sort of challenge that medical diagnosis has to offer. She lived without a diagnosis and then with the wrong one for a very long time. She had to wait nearly thirty years for a definitive diagnosis.

Stephanie's illness began in 1990, when, aged just twenty-two and twenty-nine weeks pregnant with her first child, she collapsed.

That was attributed to eclampsia, a condition manifesting as high blood pressure and seizures that occurs in pregnancy. The baby was delivered early and did not survive.

Stephanie's second collapse happened some years later, after the birth of a healthy baby girl, Abigail's older sister. That was her first convulsion. It pitched mother and child down the stairs. Both emerged unscathed but, from then on, seizures became a regular feature of Stephanie's life. She saw more than a dozen doctors and was given some contradictory opinions. First, she was told she had epilepsy. Then, when the treatment didn't work, doctors changed their minds. Somebody suggested she had psychogenic seizures caused by stress. After another emergency trip to hospital, a casualty doctor changed the diagnosis back to epilepsy. Somebody else then changed it again, back to psychogenic seizures. Some doctors treated her with medication for epilepsy, others referred her to a psychiatrist. Neither route made her any better.

I was asked to advise on Stephanie's treatment in 2007, seventeen years into her illness. As a seizure specialist, I was to put an end to the diagnostic wavering once and for all. I admitted Stephanie to hospital and simply waited to see a seizure. Most neurological diagnoses depend on clinical skills – on the interpretation of stories and on the physical examination. Just witnessing one of Stephanie's convulsions was enough. I saw the typical 'tonic' stiffening and 'clonic' muscle jerking that I had seen so many times before. It could only be explained by an epileptic convulsion. Stephanie undoubtedly had epilepsy.

But that is not to say that I could solve the mystery fully. An epileptic seizure is a symptom of many different brain diseases, including brain tumours, genetic disease, infection, inflammation, developmental abnormalities and injury. Neither Stephanie's story nor her tests helped me to narrow it down to a specific disease. Still, Stephanie was satisfied to have the diagnosis of epilepsy even if the cause for it, the *real* diagnosis, evaded us. It gave her something definite to tell people when they asked her what was wrong. It made her part of a community and opened the door to potential treatments. Not that the treatment worked particularly well. The seizures continued but, even so, having some diagnosis, albeit incomplete, felt better to Stephanie than the no man's land she had been in before. After that, all we could do was wait to see how Stephanie's illness unfolded.

To a certain degree, many difficult diagnoses are a waiting game. Two things can change an imprecise or a non-diagnosis to a definite one. Sometimes, the picture is made clearer when new symptoms emerge as the disease progresses. Preferable is that science catches up with the disease, revealing the diagnosis before the patient has become sicker. In Stephanie's case, I needed both.

A junior doctor saw it first.

'Do you think Stephanie walks strangely?' he asked me one day on the neurology ward, where I had admitted her for a reassessment.

'I don't think I've noticed anything about her walking,' I said.

But in truth, I realised I couldn't recall the last time I'd watched Stephanie walking. This was five years into my caring for Stephanie. Usually when we met, we sat opposite each other in the clinic room and talked about her seizures and sometimes about her family and her work. She had never mentioned a difficulty with walking and I had no reason to ask.

'This doctor thinks your walking is a bit off balance,' I told Stephanie. 'Have you noticed a problem?'

Her husband, Mark, who was seated in a chair by Stephanie's bed, laughed to himself.

'*He* laughs at me,' Stephanie smiled and cocked her head towards him. 'I trip over almost anything – but I've always been like that.'

I had often known Stephanie to downplay her seizures; she was not one to make much of medical problems. So I asked her more questions. She had been uncoordinated since childhood but it had become worse, she admitted. Mark agreed. The change had been so slow and ill-defined that they hadn't thought to mention it.

I watched Stephanie walk the length of the corridor. She bounced as she walked and scuffed her feet. Her toes were turned inwards. I asked her to walk an imaginary thin line, like a sobriety test. She could do it, but barely. I examined her legs as she lay on the bed. They were stiff and her reflexes were exaggerated. Some muscles were subtly weak. Although she didn't complain of any problem in her arms, the muscles there seemed abnormally stiff too.

Neurologists call this a spastic paraparesis. Spasticity refers to the stiffness of the limbs and paraparesis refers to the limb weakness. Anything that disrupts the motor nerve pathways in the brain or spinal cord can cause this. I thought about all of Stephanie's falls during her seizures. Had she injured her neck? Could her spinal cord have been damaged by trauma?

'Abigail walks like her mother,' Mark said, out of the blue.

I looked at Stephanie with surprise. When Stephanie talked about her children it was only ever to say how well they were doing. I understood them to be healthy.

'That is true, I suppose,' Stephanie confirmed. 'She's clumsy, like her mother.'

'The kids in school call her "the penguin",' Mark said.

This is the moment that still made me feel guilty years later when I finally had the chance to speak to Abigail. She was a teenager who was teased at school for having a 'funny walk'.

She was shy, but at the same time confident and capable. She liked team sports but, like her mother had been, was always picked last. She had adapted to that: swimming and yoga suited her better than anything involving running.

I jumped on the new information with the enthusiasm of a neurologist who loves putting the bits of the diagnostic puzzle together. Having ruled out a spinal cord injury as a cause for Stephanie's walking problem, I suggested that Stephanie *and* Abigail be seen by another neurologist who deals with muscle and motor problems. He agreed that both women had a spastic paraparesis which likely had a genetic cause.

The picture of Stephanie and Abigail's diagnosis came together in small leaps that happened years apart. Some genetic tests were done in 2012 but they found nothing useful. Our ability to analyse genetic codes exploded very shortly after that and, in 2019, I received a letter from Stephanie and Abigail's neurogeneticist. A new battery of tests had found a gene variant (once called a gene mutation) on one of each of Stephanie and Abigail's chromosomes. The abnormality was in the KCNA1 gene.

The disease that results from a genetic abnormality depends on what that gene does when it's healthy. The KCNA1 gene encodes for the development of a gateway that allows ions to move in and out of cells, which is essential for the normal functioning of the nervous system. Not all gene variants cause disease, but a number of different variants in the KCNA1 gene have been associated with a variety of neurological problems, including walking difficulties and epilepsy. Stephanie and Abigail's particular variant was rare. In fact, it was so rare that their specific variant had only ever been seen twice before. This genetic abnormality was newly discovered and poorly understood, but these two case reports describing other people with the same gene variant who also had neurological problems suggested we had found Stephanie's definitive diagnosis at last.

So there it was. Nearly thirty years after the fact. Stephanie had her likely diagnosis – a genetic variant in the KCNA1 gene causing epilepsy and spastic paraparesis. It was the sort of diagnosis a doctor loves. Rare. Unexpected. Hard fought. Cutting edge. It was an explanation that could potentially account for all of Stephanie's medical problems.

But a diagnosis is supposed to lead to something. Traditionally, it should explain symptoms, indicate next steps, introduce people to fellow sufferers. Stephanie and Abigail's diagnosis did none of that. They are two of a growing population who, courtesy of scientific advances, have been given a diagnosis so rare or unprecedented that nobody knows what it might mean for their future.

I didn't meet Abigail at the time of the diagnosis. I made judgements about her need to see a neurologist based on conversations with her parents that took place in her absence. I guessed that she might have the same neurological problem as her mother and assumed that was something that would be useful for her to know. She went along with what I suggested.

'I hope I didn't make you self-conscious about your body or your health,' I asked Abigail later, somewhat hopefully.

She thought for a while. 'I think I know what you're getting at,' she said. 'I see it when one of the kids at the nursery falls over. If they look down and there's blood they start screaming, but if they look and there's no blood they get up and set off again as if nothing happened.'

'I was concerned that I might have made you unduly worried about your health,' I admitted.

'No.' Abigail shook her head and smiled. 'I just got on with things.'

Eleven years after the fact, it came as a great relief to hear that.

Unlike Abigail, Stephanie had spent years waiting for a confirmed diagnosis. They were tough years but valuable too. Not knowing she had a rumbling, potentially progressive medical problem, she allowed herself ambitions and she grew her family without unduly worrying about their future. When she was a teenager, she had seen only good possibilities in front of her.

'Until my forties, I didn't know I had a genetic condition, so I just got on with things. I had a career. I had a life. When I didn't know I had the abnormal gene I could still hope it would all go away,' she told me.

'Really?! No diagnosis was better?' I asked, surprised. That is certainly not, in my experience, what most people believe.

'There is something to be said for blissful ignorance,' Stephanie smiled.

In contrast to her mother, Abigail got her diagnosis so easily, but so early. She won't go through the diagnostic uncertainty that her mother suffered, but she also hasn't had the luxury of her mother's unencumbered optimism for the future.

Abigail didn't know she had a medical problem when I recommended she see a neurologist, but it would certainly have come to light eventually. Her walking got slowly worse over time. After leaving college, she spent some time in France working as a nanny in a ski resort. By then, slippery, soft or uneven surfaces had become a challenge for her. On a slightly tipsy night out, a group of friends made name badges for each other, designed to tease. Abigail's said, *Hi, I'm Abigail, I can't walk on snow*. Now, she can walk but not for long distances and she cannot stand for long periods, either. On a recent trip to Disneyland, she needed a wheelchair for parts of the day.

If I am looking for ways that Abigail's advanced diagnosis benefitted her, I can see that she will never have to go through all her mother did to get a diagnosis. She will never feel she is not believed. Although her condition is rare and nobody can fully foresee how it will progress, she has some insight from watching her mother and seeing how things have gone for her and how well she has coped. The diagnosis has empowered Abigail to make plans. She is aware that running after the toddlers at work is getting harder and that some day she won't be able to, or won't want to, do it anymore. She has started to look for a more suitable long-term career that will not be so physically taxing.

Having a diagnosis has given her access to practical support, such as disabled parking. She can bring a carer to venues and events that might be physically challenging for her. If she ever needs it, a formal diagnosis will simplify her access to financial support. She now knows that a genetic condition runs through the family. She may be able to avail herself of technology that will allow her to have a child who is free of her genetic disease – or not, if that's what she prefers. She has a choice because, limited as it is, she has a diagnosis.

But it could have gone a different way. Being told she had a spastic paraparesis could have made Abigail pay undue attention to her legs. Instead of playing team sports and accepting that she would never be the strongest player, she could have avoided doing anything that she thought she wouldn't do well. Instead of becoming a nursery nurse, she might have abandoned her first choice of career in favour of something less physically demanding. She might even have been turned away from a job on medical grounds if a new employer had learned of her diagnosis. She could have been denied insurance on the basis she had a progressive medical condition.

Abigail is resilient and pragmatic like her mother, so she has been able to put the diagnosis to the back of her mind and get on with life – but I didn't know that about her when I sent her to the neurology clinic. Another person might have incorporated the diagnosis into their identity and changed their lives for it. The word patient is drawn from the Latin verb *pati,* which means to suffer. When I made Abigail a patient, I could have made her suffer. That I didn't was only my good luck.

[illegible]

Ever since I asked [illegible] she has taken to [illegible] in her spare time [illegible] and birds. They arrive [illegible] make me smile. I can tell [illegible] perhaps [illegible] made Stephanie feel better [illegible] I have failed to eradicate her seizures [illegible] value of the diagnosis I gave her [illegible] done with a cure or [illegible] that will lessen symptoms [illegible] as having an explanation [illegible] diagnostic label?

I have now been a [illegible] for twenty-five [illegible] my patients in mind [illegible] worried for the large number of [illegible] three, four or five [illegible] only some of which [illegible] ADHD, migraine, fibromyalgia [illegible] depression, eating disorders [illegible] many new diagnoses [illegible] but which [illegible] Danlos syndrome, [illegible]

Introduction

Ever since I asked for permission to tell Stephanie's story, she has taken to sending me photographs of paintings she does in her spare time. Cheerful watercolours depicting flowers and birds. They arrive by email with some regularity and always make me smile. I can tell that something about having a diagnosis or, perhaps, something in our shared journey together, has made Stephanie feel better, however resolutely I have failed to eradicate her seizures. And yet, I still feel ambivalent about the value of the diagnosis I gave her. Or any diagnosis that does not come with a cure or at least a moderately effective treatment that will lessen symptoms. Beyond the initial moment of relief at having an explanation, what are the other implications of a diagnostic label?

I have now been a doctor for over thirty years and a neurologist for twenty-five of those. When I write a book it is always with my patients in mind. In recent times, I have grown particularly worried for the large number of young people referred to me with three, four or five pre-existing diagnoses of chronic conditions, only some of which can be cured. Autism, Tourette's syndrome, ADHD, migraine, fibromyalgia, polycystic ovarian syndrome, depression, eating disorders, anxiety and many more. There are so many new diagnoses that did not exist when I was a medical student but which are now common. Hypermobile Ehlers-Danlos syndrome, postural orthostatic tachycardia syndrome and many novel genetic disorders. Where did they come from and how did they go from zero to a hundred so rapidly? I am constantly shocked that so many people in their twenties and thirties could have accrued so many disease labels at such a young age. Older people, too. Hypertension, high cholesterol, low back pain and so on. It is becoming unusual for me to meet a patient who does not have a trail of prior diagnoses. They are sent to me with a new symptom and they hope that I can provide another diagnosis to explain that. Medicine has many blind spots in which practices become absolutely routine without anybody noticing that they aren't solving the problems they promised to. I have been wondering for a while if the long lists of diagnoses I am seeing are part of this phenomenon.

A person does not have to be a medical professional to have noticed that certain diagnoses have become suddenly common. The startling rise in people diagnosed with mental health disorders, behavioural and learning difficulties features regularly in newspaper headlines and in our conversations: 'ADHD, what's behind the recent explosion in diagnoses?' *New Scientist*, May 2023.[1] 'Autism prevalence rises again, study finds', *New York Times*, March 2023.[2] The story is the same for multiple categories of mental health disorder. 'PTSD has surged among college students', *New York Times*, May 2024.[3] 'More than 1 in 6 adults have depression as rates rise to record levels in the US', CNN, May 2023.[4] 'Depression and anxiety rates have increased by 25% in the last year', *Forbes*, February 2023.[5]

But it's not just mental health and related conditions that are on the rise. Asthma diagnoses increased by 48% in the US in the last twenty-five years.[6] The number of people with cancer was projected to exceed 2 million in the US for the first time in 2024[7] and dementia diagnoses in the UK reached a record high.[8] Worldwide, 537 million people are living with diabetes, with 783 million predicted to be affected by 2045.[9] The rate at which people are being found to have all the following physical health conditions has increased steeply in the last twenty years: cancer, genetic diseases, dementia, hypertension, hypercholesterolaemia, diabetes, osteoporosis, kidney disease, polycystic ovarian syndrome, endometriosis, pulmonary emboli, aortic aneurysms, chronic Lyme disease. And so many more besides.

What do these astonishing statistics say about the state of our health? On the surface, they make it seem as if we are considerably less mentally and physically healthy than we used to be. But there are other ways of interpreting them. Could they simply reflect the fact that we are much better at recognising medical problems and identifying people in need of treatment? Disorders like autism may be on the rise because people are finally getting the right diagnosis and being given support. Perhaps more of us are aware that we have hypertension and diabetes because our doctors are more proactive about looking for them and treating them. If that's the case, it could be that more diagnosis actually indicates that we are becoming healthier.

But there is a third possibility. It could be that not all these new diagnoses are entirely what they seem. It could be that borderline medical problems are becoming ironclad diagnoses and that normal differences are being pathologised. These statistics could indicate that ordinary life experiences, bodily imperfections, sadness and social anxiety are being subsumed into the category of medical disorder. In other words: we are not getting sicker – we are attributing more to sickness.

Which of these explanations is most likely to be correct is a matter on which it is very hard to find agreement. It is subject to widespread debate between medical professionals and the public alike. But it is very much in our interest to find the answer because the trend to detect health issues in milder and earlier forms, the assumption that is always the right thing to do, is pressing forward relentlessly. In this book, I will make a case for possibility number three – that we are becoming victims of too much medicine and that it is time to turn back the dial. That we are living in the age of diagnosis for both mental and physical health conditions.

Developments in medicine and changes in society have led to two rapidly increasing phenomena, which I will examine in detail in this book: overdiagnosis and overmedicalisation. While misdiagnosis simply means getting the diagnosis wrong, overdiagnosis is something entirely different. It refers to a diagnosis that is correct but which does not benefit the patient and may arguably do harm. Overdiagnosis occurs when a medical problem is detected at a stage when medical treatment is not really required. Such as telling fifteen-year-old Abigail that she had an untreatable genetic spastic paraparesis before she was aware there was something wrong. Such as too much health screening of asymptomatic individuals with no evidence it leads to longer or better lives. Such as too much aggressive treatment for early disease. Such as too much unnecessary health monitoring.

Overmedicalisation is related, but subtly different, to overdiagnosis. It occurs when ordinary human differences, behaviour and life stages are given medical labels, turning them into the business of doctors. Like telling immature or socially anxious children that they have a neurodevelopmental brain disorder. Or turning non-diseases into diseases with an expectation that they are problems that medicine should cure, as we are seeing with ageing, poor sleep, sex drive difficulties, menopause and unhappiness.

There are two principal mechanisms through which overdiagnosis and overmedicalisation arise. The first of those is overdetection. This occurs when new technologies and more sensitive and intensive screening programmes are used to detect earlier and milder forms of disease. The second is expanded disease definitions. This occurs when the dividing line between normal and abnormal slowly shifts so that, over time, people who would once have been considered healthy are drawn into the disease group. This is also called 'diagnosis creep'.

Overdiagnosis and overmedicalisation usually arise out of good intentions, but also out of presumed truths that have not been tested. Such as the assumption that early diagnosis is always for the best. That having a diagnosis is always better than no diagnosis. That people want to know their medical future even if it can't be changed. That more medicine is better medicine. That modern medicine is better medicine. That high tech is better than low tech. But medical change is supposed to be based on evidence, not on assumptions.

Everybody is rightly concerned about underdiagnosis. Many of us have at least one personal story of an illness that was missed by a medical professional. Overdiagnosis is much harder to spot and therefore far less talked about. Someone who is told that they have a disease or are at risk of a disease is rarely in a position to disagree and reject their doctor's conclusion. It is very hard for an individual to determine that they have been given treatment that was unnecessary, meaning you will hear few people complain about being overdiagnosed. It is also very difficult to identify the point at which appropriate medical treatment becomes overused or diagnosis turns into overdiagnosis. For this reason, and because we are so afraid of missed diagnoses, it has been very easy for us to stumble in the direction of too many diagnoses made too easily.

There is a growing body of evidence that overdiagnosis might even outstrip underdiagnosis and that it comes with its own set of harms. Consider these scenarios. NHS England estimates that treating cancers found in health screening programmes saves 10,000 lives per year.[10] But *what if* those very early cancer cells found on screening were never destined to cause serious illness? Not all early cancer cells progress to full-blown cancer. Perhaps, among this 10,000, some lives were saved but other people were given cancer treatment that wasn't needed. This is actually a very common example of overdiagnosis by overdetection. A 2023 study carried out in the US estimated that 31% of breast cancers diagnosed in women over seventy were overdiagnosed.[11] A French study estimated that more than €100million was spent on overdiagnosed thyroid cancer in a four-year period.[12] In many prostate cancer screening programmes *no* lives are saved but, for every thousand men screened, as many as twenty men are diagnosed with and treated for cancer that would never have caused problems if left alone.[13] This story is repeated worldwide, for every type of cancer that is subject to screening. Cancer screening programmes save lives but they also risk exposing people to unnecessary invasive medical treatment and to the psychological drawbacks of a cancer diagnosis.

Expanded disease definitions – this moving of the goalposts that means more people are included in the group considered to have the disease – have had a dramatic effect on the rate of diagnosis of lots of medical problems. Consider the case of pre-diabetes – a stage between entirely normal glucose levels and actual diabetes. In 2003, the American Diabetes Association adjusted the definition of what it means to have pre-diabetes, lowering the threshold for a normal glucose level in a fasting person from 6.1 millimoles per litre to 5.6.[14] That seemingly small change led to at least a two-to-threefold increase in the number of people with pre-diabetes overnight. Were this lowered bar for qualifying as pre-diabetic combined with other tests of glucose intolerance it could mean as many as half of Chinese adults and a third of British and American adults would be considered to be pre-diabetic if the changes were implemented fully at a global level. That would place them in a group thought to be at high risk for developing diabetes and make them subject to medical monitoring and potential health anxiety.

This abrupt escalation in diagnoses of the condition arose out of nothing more than a committee decision to redefine the parameters of normal glucose. The idea was that all these extra people with the pre-diagnosis would get treatment or advice that would help them to avoid developing diabetes and enjoy better long-term health. But it is unclear whether this change did actually lower rates of diabetes as it had promised to do. Global type 2 diabetes prevalence rises year on year despite these measures having come in over twenty years ago. Diagnosing pre-diabetes has been shown to delay the onset of diabetes in some people to their benefit. However many, and some would even say most, people regarded as pre-diabetic would never have developed diabetes if left undiagnosed so, for that group, the monitoring may never have been needed. Have more inclusive diagnoses that draw in milder cases saved lives or medicalised them with little gained?

Now let's think about all those new people who have recently learned they have autism and ADHD. Identifying struggling children and adults and explaining those struggles through these diagnoses is supposed to lead to happier, more successful people. We know that intervention for these disorders, if given promptly to young children and to those more severely affected, does help their progression through life. But recent changes to the definitions of these conditions have seen many more people with much milder 'symptoms' and older people being diagnosed. There is considerably less evidence that social or medical interventions work in this group. Is everyone who receives these diagnoses really benefitting from them? Overdiagnosis does not say the diagnosis is *wrong* but rather points to the potential that harm outweighs benefits. Could there be untold harms in telling people they have a brain disorder that cannot be offset by treatment? After the momentary relief that comes with an explanation has passed, what next? Diagnosis may have a negative effect on psychological wellbeing and on social standing that is hard to measure.

This is the biggest marker of overdiagnosis – much higher rates of detection of diseases but no substantial improvement in long-term health. All too often, the assumption that better, earlier, more advanced, more inclusive diagnosis is definitely for the best is so strongly held that it is never properly tested. My worry is that the scientific community's hunger to use all its new diagnostic capabilities to their fullest capacity and to find medical problems in their mildest and earliest forms – along with our natural, human thirst for explanations – has not allowed enough time to weigh up value against harm.

Importantly, overmedicalisation and overdiagnosis are not just things that the medical community are doing to the unsuspecting public. The pathologising of distress, the sanitising of the messy truth of life through biology, is a scientific *and* a social trend. If we do indeed have an overdiagnosis crisis, how is society's expectation of success and perfection playing into that? We are encouraged to believe anything we want should be possible for us, but that cannot always be true. I fear that medical diagnosis has become something that is used to reframe our perceived failings. An unrealistic expectation of an unachievable level of physical and psychological perfection might be turning us into patients and robbing us of control over our own destiny.

As a neurologist, much of my current work is spent looking after people with brain diseases like epilepsy. But I also care for another large group of people whose experiences are very pertinent to the subject of this book. They have a psychosomatic disorder meaning they have very *real* physical symptoms that happen for psychological reasons. We are all vulnerable to overdiagnosis but this group is particularly vulnerable. In an atmosphere in which there seems to be a diagnosis to explain every human experience, people naturally prone to notice and worry about ordinary bodily changes, people in need of reassurance, are increasingly given a diagnosis instead. Now that we have a disorder label for almost every sort of physical variation and all levels of mental anguish, I fear that people who express their emotional distress as physical symptoms can all too easily have them conflated with disease.

Psychosomatic disorders often arise out of the stories we tell ourselves about our health. Telling a person they are sick – advising them of developmental problems and chemical imbalances and impending disease – changes how they perceive and use their body. Abigail put it beautifully when she described how a child's reaction to a fall differed depending on whether or not they looked down and saw blood. Sometimes, a disease label has the same effect as a person, pointing at your body and telling you there is blood. It can alter the entire experience of your body. Labels have the power to actually make us sick, through a mechanism called the 'nocebo' effect, which we will soon explore. When we don't feel well, it's natural to want answers. Doctors and patients work together to find that diagnostic label because it feels good to both of us. But perhaps people would be less keen to agree to be labelled if they understood that, for some, a diagnosis can create an expectation of illness that actually generates symptoms even when there is little or no disease.

But the potential psychological harms of overdiagnosis are not only for those prone to psychosomatic disorders. Think of that 31% of women over seventy in the US who could be overdiagnosed with breast cancer on screening. How will the later years of their lives be overshadowed by that unnecessary diagnosis? Think of all those previously healthy people told they are pre-diabetic and all the blood tests and appointments with doctors that now punctuate their lives.

Nor are the harms of overdiagnosis only for the individuals involved, but also for the wider community with that diagnosis.

As the concept of what it means to be depressed or autistic changes to include those with much milder symptoms, how does that impact those with the severest forms of these conditions? As an increasing number of people are labelled hypertensive or pre-diabetic, as more is said about the negative effects of the menopause or the impact of poor sleep, how are those with the most severe manifestations of these problems affected by this shift? When the number of people with a diagnosis rises, it may lead to better services and better public knowledge about and sympathy for the condition. But it is equally possible that a growing population with the mildest form of any condition risks trivialising the disorder for those most severely affected and diverts resources from those most in need.

It doesn't help that in many medical fields, our ability to diagnose outstrips our ability to cure, meaning people are living with a diagnosis for longer, but not necessarily living longer as a result. Early dementia diagnosis may be at a record high in the UK but there is still no treatment that will stop the disease from progressing. The fate of a person given an early diagnosis of dementia cannot be changed. Not yet, anyway. So too for many newly identified genetic disorders. The first ever genetic variant known to cause disease was found in 1993 and now we know of *millions* of others, each providing a novel diagnosis for a new set of people. Like Stephanie and Abigail. Traditionally, a diagnosis is supposed to lead to treatment and inform us about prognosis, but these new genetic diagnoses rarely do any of that. That we are changed by the labels given to us is particularly a problem given how many medical labels and early diagnoses are now available.

In writing this book, I explore how modern medicine is redrawing the boundaries between sickness and health and what impact it is having on our lives. I have grown very concerned that we are turning increasing numbers of healthy people into patients with insufficient gain and unmeasured harms. Diagnosis is seen as the key that opens the door to so much that we long for – an explanation, the potential for recovery, support, a tribe of fellow sufferers – but there are darker things behind that door too that are not always given proper consideration.

*

In this book, I will challenge many commonly held assumptions – that any diagnosis is better than no diagnosis; that tests are more accurate than doctors; that test results are objective, immutable truths; that early intervention is always for the best; that treatments that work for one set of people will work equally well for others; that diagnosis is something fixed and definite; that pre-emptive testing is the surest way to long-term health; that more knowledge is always for the best.

I will also examine how even the concept of diagnosis is changing for the better and for the worse. A growing acceptance of self-diagnosis is shaping medical problems independently of scientific discovery, leading to a startling impact on some patients. Autism and ADHD self-diagnosis questionnaires are now widely available on the internet and self-diagnosed people are entering research studies alongside the formally diagnosed. The development of new diagnostic concepts was once the domain of doctors, but that is no longer the case. How is patient-driven research and social pressure changing the shape of diagnosis? Long Covid was patient-created – the first disease to arise from a thread of conversation on Twitter. I'm sure it won't be the last.

I will try to understand what is driving the uptick in so many different types of diagnosis. Some of it is surely coming from doctors and scientists who are keen to use all the new technologies at their disposal. But we can't lay all of this at their door. Many of us are searching for explanations for all sorts of physical suffering and personal struggles. We are demanding answers and, in the absence of other sources of support, are looking to medical institutions for help.

Each chapter will examine a different type of diagnosis and include stories of real people who have been the beneficiaries of all that modern diagnostic medicine has to offer. I could not hope to cover every kind of mental and physical diagnosis in this book, so I have chosen subjects which represent a different emerging theme in diagnosis and where I think there are broader learnings.

I begin with Huntington's disease, a disorder that most of us will never encounter but which has an incredibly important story to tell that will matter to everyone very soon. Decades before anyone else, the Huntington's disease community was able to avail themselves of the advanced diagnosis of a disease in their distant future. As tests become more sensitive and genetic diagnosis becomes more available, more of us might soon be offered the same. If you were destined to develop dementia in ten years' time and it remains incurable, would you want to know? This chapter will examine how it feels to live with knowledge like that and will challenge many people's assumption that knowing is always better than not knowing.

Next we'll look at Lyme disease and long Covid – disorders that have touched many people's lives. They have a great deal in common. Both started in a very unconventional way – with a patient-led campaign – and both have become shrouded in controversy. But the lesson they teach is transferable to all diagnosis. Tests are not as accurate as we think. They may even pretend to offer accuracy while actually contributing to error. Diagnosis is subjective, a true art, and that makes it susceptible to mistakes, exploitation and social pressure.

In a chapter about autism, I will consider how something that used to be chronically underdiagnosed has become so common and ask why those with the condition now look so different to the original autistic children of the 1940s. I will look at how diagnoses naturally evolve and grow over time not based on scientific advances but on social consensus. I will ask if the autistic community, both as individuals and as a group, are benefitting from this more inclusive version of the disorder.

Cancer is on the rise. Screening for cancer and searching for people at risk saves lives. In a chapter centred around the stories of women who learned they had a cancer gene, I will ask how reliable advanced and early diagnoses really are. Might we be so enamoured with our new technical capabilities and our enthusiasm for rooting out early disease that we are compelling people into medical treatment that they never actually needed? This chapter will also explore some of the more troubling consequences of the huge rise of over-the-counter genetics tests produced by private companies.

A popular conversation these days is to wonder if we are pathologising normal. Many fear that mental struggles that might once have been referred to as ordinary are being reframed as medical and spoken of in biological terms. People are not sad, they have low serotonin levels. They are not forgetful, or fidgety, or unsettled, their brain is wired wrongly. In a chapter on neurodiversity, I will wonder if every mental health diagnosis is a medical problem and will ask if the emotional payoff that comes with such a diagnosis could get in the way of recovery. What happens to a person when their illness becomes integral to their identity?

Often, in talking to the people featured in this book, I had the sense that too many of us are demanding perfection of ourselves. But some of us are demanding it of our children too, and of our unborn children. In a chapter entitled 'Syndrome without a Name', I will look at the ethical and practical issues of making advanced diagnoses in children, babies and the unborn. Childhood is a time when a person should be allowed to believe that any type of future is possible. Is predicting future diagnoses in children and choosing genetically 'perfect' babies really the best way to create healthier, happier future generations?

In the conclusion, I will come back to my own patients and the reason I wanted to write this book. The last thirty years has seen the rise of new diagnoses, like hypermobile Ehlers-Danlos syndrome and postural orthostatic tachycardia syndrome. If you haven't heard of them yet, you will soon. Where did they come from and are they used to pathologise ordinary changes in growing bodies? We are systematically developing labels to explain every imperfection and bodily difference. It is becoming difficult for anybody at any age to consider themselves perfectly healthy.

This is not a book about any particular health service, in the UK or the US or anywhere – overdiagnosis is a global issue. In writing *The Age of Diagnosis*, I've drawn on my own clinical practice but have also spoken to dozens of patients, as well as medical specialists and researchers around the world. I've dug through medical archives to unearth how new diagnoses come about. I have listened to the heart-wrenching stories of healthy people advised that at some indeterminate point in the future they might die of cancer unless they take drastic, life-changing steps to avoid it. They have made an almost impossible choice but, in doing so, saved their own lives. I have talked to parents given ultra-rare genetic diagnoses for their children and to young people who have been told that their lives will be foreshortened by untreatable degenerative conditions. They taught me that the value of a life does not lie in how long it lasts or in flawless health, but in relationships, in small pleasures and very personal achievements. I have spoken to people with autism and ADHD, to parents and teachers of neurodivergent pupils, in the hope of getting a better understanding of what stands to be lost and gained when a person's feelings, behaviour, their way of being are labelled as medically abnormal. I have tracked the course of disorders like Lyme disease and long Covid from their very beginning, to get a sense of how new diagnoses form and to try to understand how public activism and social media is likely to change diagnosis in the future. Most people I spoke to valued their diagnosis, were buoyed by it, but they also had strong feelings about what should change.

Neither is this a book about saving money for health services. You might think that better diagnostic prediction tools, pre-emptive diagnosis and better recognition of mental health problems are a cost-efficient way of creating a healthier population. That is certainly one of their purposes. But more people getting a diagnosis at earlier stages can be very expensive. Overdiagnosis is low-value care. It means spending money to treat people for diseases that would never have progressed and monitoring people with minor symptoms that would have resolved spontaneously if left alone. But that does not mean this book is about rationing medicine to make it more cost-effective. It's about a search for a better balance between the benefits of our diagnosis-driven culture and a recognition of the physical and psychological downsides that come with it. It is a search for better medical practice.

Society does not have a very good history of how it manages its new technical and scientific capabilities. Antibiotics, opioid analgesics, plastics, fossil fuels – when society invents or uncovers something that is transformative, it has a tendency to celebrate it, overuse it and misuse it. And often, it is not until the abuse of our capabilities has gone too far that we recognise our mistakes. This seems a good time to cast a quizzical eye over the gift and burden that is modern diagnosis. We need to make sure we have the balance right. Our new capabilities and attitudes are very seductive. If there is a singular, tangible medical explanation that explains how we feel, we welcome it. When scientists and doctors are capable of doing something clever, it often feels that we automatically do it every time. But just because we can, doesn't necessarily mean we should.

I
Huntington's Disease

'If you were going to be knocked down by a bus tomorrow, you wouldn't want to know, would you?' Valentina asked.

'I *would* want to know,' I decided. 'I'd spend a lot of time talking to people and then I'd eat everything off the menu in the Ritz!'

We were both laughing but we were also aware that Valentina knew better. Valentina's mother has the genetically determined neurological condition Huntington's disease (HD) and Valentina faced a 50% chance that she had inherited the disorder. She had spent years considering whether or not to be tested, weighing all the pros and cons of having advanced knowledge of a big slice of her medical future. A predictive genetic test could give her some sense of certainty – but, like all predictive diagnostic tests, it would also create a cascade of consequences that many fail to anticipate.

Predictive medicine is diagnosis in healthy people before a disease has had a chance to start. Genetic predictive diagnosis can be used to advise people of a pending health problem decades ahead of time. That means many years of waiting and watching for the first symptom to strike.

★

Huntington's disease is an incurable condition that causes progressive physical and cognitive disability. The early signs are often subtle behavioural changes such as mood swings, social withdrawal, loss of impulse control and poor organisation. Psychiatric features are prominent. Depression, mania, obsessive behaviour and thoughts of death and suicide may feature. Motor symptoms can be present early on but more often come later, resulting in clumsiness, loss of balance, slurred speech and difficulty swallowing. A hallmark of the disease is the development of jerky involuntary muscle twitches called chorea or choreiform movements. In the later stages, sufferers are unable to walk and struggle to eat or speak. Nothing will stop the downhill trajectory to severe disability. The typical age of onset of symptoms is between thirty and fifty, with the eventual death coming roughly ten to twenty-five years later. The first disease gene to be discovered was that for Huntington's disease. More than anyone else, the HD community understand the gravity of the decision to take a predictive diagnostic test.

Human DNA is arranged into twenty-three chromosome pairs. One pair are sex chromosomes, X and Y, and the other twenty-two numbered pairs are referred to as autosomal. Genes, which make up small sections of the long threads of DNA within a chromosome, are the basic unit of inheritance. They contain all the instructions needed for human development. Genes make proteins which make cells, and we are made of cells. When a gene code contains some form of mistake, a genetic disease can develop. Mistakes in genes were once called mutations but are now called variants.

As soon as the normal structure of DNA was understood the search for genetic variants that cause disease could begin. The precise faulty gene that causes HD was identified in 1993. It is a monogenic disorder, meaning it's caused by a mistake in only one gene. Polygenic disorders, on the other hand, have contributions from multiple genes. HD is autosomal dominant, so the disease gene is on a numbered autosomal chromosome – chromosome 4 – and only one gene in a pair needs to be a disease gene for HD to develop. Monogenic dominant disorders like HD have predictable inheritance because children either get the genetic variant from their affected parent or they don't: a fifty-fifty chance. The HD gene variant is not just a risk factor suggesting that a person *might* get HD, it is a certainty. The only question is when the symptoms will start and how quickly they will progress.

The discovery of the HD variant brought testing for Huntington's disease into everyday medical practice. From that moment on, any person with a family history of HD could be tested to see if they were destined to develop the condition. It was clinical medicine's first opportunity to grapple with all the implications of giving healthy people a diagnosis of an impending incurable disease.

⋆

Valentina was twenty-eight and pregnant with her first child when she learned her mother was being tested for Huntington's disease. Valentina's family had no idea this was on their horizon. Vivian, Valentina's mother, was adopted. She was never told of any bleak medical secret in her biological family, so when, in her fifties, she started dropping things and developed tic-like muscle twitches and loss of balance, nobody recognised this as the start of HD. Much later, Valentina wondered if her mother's symptoms actually began many years before. As Valentina remembers it, Vivian had been prone to depressive bouts for many years. She was easily flustered and bad at decision-making. These are exactly the sort of non-specific symptoms that herald the insidious onset of neurodegeneration. When they happened to Vivian, they were attributed to more ordinary mental health problems.

A brain scan was what first alerted Vivian's neurologist to the possibility of HD and the need to refer Vivian for a genetic test. Vivian didn't take the suggestion very seriously. She was sure that somebody in the adoption services would have told her that she had a parent with HD and because they didn't, she convinced herself it was not her diagnosis. Only later did the family learn that Vivian's father had spent years in a psychiatric institution. In the past, before HD was fully understood or easy to diagnose, some sufferers languished in psychiatric units without a diagnosis. Not knowing she was at risk, Vivian was absolutely certain her tests for HD would come back negative. She was so dismissive of the test that Valentina wasn't particularly concerned either and didn't even think to ask about the result.

The day that Vivian found out she had HD, Valentina's future and the future of her unborn child changed dramatically. All of Valentina's siblings' and nieces' and nephews' lives changed, too. Each of Vivian's four children went from being perfectly healthy young people to having a fifty-fifty chance that they would eventually develop an incurable neurodegenerative disease. Vivian's grandchildren had an immediate 25% chance of being affected. The wife of Valentina's brother Luca was pregnant too, with their second child. Valentina's younger sister Camila was about to get married and had yet to start a family. The eldest of the siblings, Evangeline, already had three daughters and a son.

Vivian waited until Valentina's baby daughter, Ella, was born before she told her that the genetic test had been positive for HD. It came as a complete shock.

'I knew exactly what it meant when I heard about the diagnosis,' Valentina said, thinking back to that time. 'I like medical programmes and books. I'd heard of HD and I knew it was bad.'

Vivian's symptoms had probably begun in her early forties. If Valentina had inherited the gene variant that causes HD, she would most likely start showing signs of the disease around the same age as her mother, giving her maybe another fifteen years of good health.

Vivian's diagnosis had an immediate impact on Valentina. She had always been a light-hearted, glass-half-full type of person, but she changed. She began having panic attacks. Before long, she needed antidepressants to control her mood and she has continued taking them since.

'When this is in the family, the children need to be prepared from a young age,' Valentina advised me. 'For us, there was no preparation and we all had so many decisions to make and so quickly.'

Valentina and her siblings were each at the point of starting or expanding their families. They were developing their careers, buying houses. Should they keep their plans the same or change them to accommodate the new possibility? Of the four children, surely at least one, probably two, would have inherited their mother's disorder.

'Did you think about getting tested straight away?' I asked.

'I thought about it, but hardly at all. I had a baby to look after and that was my priority. My brother Luca was the one who insisted he wanted to be tested immediately. But he didn't do it in the end. He still hasn't.'

Valentina and I were speaking twenty-two years after she learned of her mother's diagnosis. Valentina had waited twenty years to have the test that would let her know if her mother's fate was also her own.

'What stopped you testing for so long?' I asked. My first impulse as I began exploring predictive genetic diagnosis for this book was to assume I would take the test at the first opportunity – just as I was sure that I would want to know about the metaphorical bus that would soon knock me down. Valentina and her siblings did have an immediate thought to test but, ultimately, they all put it off. I wondered what changed their minds.

'Hope,' Valentina answered. 'If you don't test, you can hope you don't have it and you cling to that. You don't want to know because you want to have a free life. Hope will carry you a long way.'

Nonetheless, Valentina's life was shattered by her mother's diagnosis. She sought the support of a psychologist. It didn't help. Valentina's worry was too specific and the non-specialist counsellor didn't understand the decision Valentina was facing. Only when Valentina found someone who specialised as a genetic counsellor did she get some peace of mind. Genetic counsellors are not psychologists. They are experts in genomic medicine, skilled in calculating genetic risk, explaining inheritance patterns, predicting risks of genetic disease, advising on genetic testing and, crucially, helping people to understand and live with uncertainty. Valentina's genetic counsellor was the first person who fully grasped the complexity of her decision. They challenged her impulse to test, helping her realise she wasn't ready. She was a young mum. She would prefer to get on with her life. Valentina, supported by her husband, found ways to manage the worst of her panic. Although, even on the happiest of days, anxiety always lingered in the back of her mind.

There were also big decisions to be made. Valentina and her husband, Jonathan, never doubted they would have more than one child. Her brother Luca also wanted more children. Camila, the youngest, had not even started her family yet. But choosing to have a child naturally, who would automatically have a 25% chance of developing an incurable condition that foreshortens life, was not easy for any of them.

One way that predictive genetic diagnosis has certainly benefitted people is that it offers a means by which families with a monogenic condition can dramatically lower the chance of passing it on. Pre-implantation genetic testing (PGT) is used to check embryos for genetic mistakes. It means having a child by IVF. Only those embryos with a normal chromosome 4 are implanted. At the time that Valentina and her husband were making the decision to have their second child, PGT cost £30,000. The NHS now provides it for free to people with a family history of select monogenic disorders. PGT, like any IVF procedure, is much less likely to result in a successful pregnancy than natural conception. It can also be a gruelling process for the prospective mother.

Valentina and her husband agonised over the decision. They were counselled about the pros and cons of PGT. They decided they would prefer to have a child naturally.

'It wasn't the money,' Valentina told me. 'Or it wasn't *just* the money. I couldn't imagine sitting my children down one day and telling one child that they were OK, that they'd been chosen. Then having to tell Ella that she had a fifty-fifty chance of HD. I never wanted to put my children in that situation. It was a conversation I knew I couldn't have.'

Valentina felt a lot of guilt about deciding to have a child who could later develop HD. But it is actually a common decision for people in her position to make. Her brother Luca did the same. They wanted all their children to feel equal. Camila, who had no children when she learned of her mother's diagnosis, chose to have PGT.

Valentina gave birth to baby Jake, five years after his older sister, Ella. Valentina was thirty-three and still hadn't tested.

During all these hard decisions, the siblings watched their mum become significantly more symptomatic. It was painful to see. Vivian was often aggressive and argumentative. She had violent outbursts. Verbal assaults on Valentina's father evolved into physical attacks. Valentina, who lived nearby, was often called in to help calm things down. Some fights were so furious that the police were called. Vivian was abusive to complete strangers. If she encountered a person who was mildly overweight she called them 'f**king fat'. She made racist comments that were entirely out of character. Her mood was very low. She once tried to throw herself out of an upstairs window. These are all extremely common behaviours for someone with advancing HD.

Vivian's physical disability also accumulated slowly. She developed chorea – constant, unpredictable, fidgety movements. Her balance deteriorated and so did her speech. Ultimately, the family had no choice but to move her into a nursing home for specialised care.

'I started to dread visiting her. I was seeing my future and my kids' future,' Valentina told me. 'It was very difficult because I adore my mum. We're very close. It was hard to see her get worse for her sake, but she was also a constant reminder of what might be ahead.'

As Valentina watched her mum go downhill, she became more and more anxious, particularly once she began to detect symptoms in herself. Her sister Evangeline, four years older than Valentina, had noticed the same. They were clumsy. They dropped things. Their moods were changeable. The sisters spent hours on the phone talking about symptoms. They tried to reassure each other. Valentina insisted that Evangeline was mistaking premenstrual symptoms for HD. In one of those conversations, Evangeline got annoyed with Valentina. They argued. Evangeline told Valentina to stop giving her false reassurance because she knew she had it. She just knew.

Evangeline was soon proven right. Six years ago, she tested positive for the HD gene. By then, her four children were in their late teens and early twenties. With Evangeline's positive test, her children's chance of having HD jumped from 25% to 50%.

Valentina was devastated for her beloved sister, to whom she had always been close. And it didn't help that Valentina also had all the symptoms that Evangeline and her mother had experienced. She got flustered if she had to do two things at once. She veered to one side. She walked into walls. She had mood swings just like her mother. Sometimes, Valentina didn't want to go out because she felt so dizzy and off balance, and she feared it had become noticeable to others. She travelled a lot for work and her symptoms were at their worst in airports. The precise timings and documents involved in catching a flight were too much for her. Just approaching check-in she became flustered, which would spiral into a feeling that she couldn't cope. She dreaded trips abroad. As her symptoms accumulated, Valentina still agonised over the decision to test. Without the positive test she still had hope, but she knew she couldn't put it off forever.

Genetic tests are not usually done casually because they come with so many personal and family implications. In the UK, a minimum of three genetic counselling sessions are needed before a person can have a predictive genetic test for an incurable condition. When Valentina's anxiety peaked or she had a new symptom, she'd book some counselling sessions with the geneticist. She'd go to the appointments with the intention to test but each time the counsellor would say something that would make Valentina realise she still wasn't ready. Despite her anxiety, Valentina usually managed to be her previous upbeat, happy self with her children. She'd had a fun childhood and that was the childhood she wanted for them. She didn't think she could maintain that if she was confirmed as having Huntington's. Not knowing allowed her to keep pretending. She wanted her children to have a happy mother.

'If I tested positive I was afraid I would look at my children, my beautiful children, and all I would see in their future was HD.'

The counsellor kept reminding Valentina that not every change in mood and not every new dizzy spell was necessarily due to HD. Valentina was not the only person to hate airports or to suffer anxiety when travelling. Her symptoms were not unusual enough that they couldn't have another explanation. If she tested positive for HD, she might stop looking for other explanations and stop trying to fix things that might be fixable.

When Valentina finally had the test, decades after her mum's diagnosis, it was because she was so symptomatic that it was ruining her life. She could no longer avoid getting the confirmation she dreaded. Ella was nineteen and Jake fourteen. At first, Valentina and her siblings had hidden their mother's HD diagnosis from their children and their friends. Their greatest fear was that the children would find out by accident and that they would grow up afraid. But the genetic counsellor helped Valentina to realise that normalising the disease in the family and familiarising the children with it when they were young would mean they would never face the shock that she had. So she had been drip-feeding information about HD to Ella and Jake for years. For a while, she just said that granny had a neurodegenerative disease but she didn't name it. HD information leaflets lay around the house. When the children were old enough, she sat them down and explained that granny had HD and that she might have it too. Valentina was shocked to learn that Ella already knew. HD is often used to teach genetic inheritance in schools and she had studied it there and had guessed what was wrong with her grandmother.

'That upset me too,' Valentina said, 'that she knew all along and didn't come to me.'

But Ella hadn't spoken to her mother about it simply because she wasn't worried. Her parents had always told her that granny's disease was hereditary and had emphasised all the positive research. That was good enough for Ella. Jake, who was younger, had never heard of HD but took the news on the chin.

A furious argument with her husband was the trigger that finally forced Valentina to test. The degree of anger that Valentina displayed was out of character for her. She had locked herself in their bedroom in a fury. There was something very familiar and worrying about the tone of the argument. She had seen many similar rows between her parents in the early stages of her mother's illness. Valentina finally told her children that she was going to have the test.

Ella made her feel better. 'I want you to have the test, Mum. You'll be happier if you know.'

It took several months for the process to be completed. Although Valentina had had counselling on an intermittent basis for many years, she still had to have the minimum three counselling sessions again before she could have the test. The sessions had to be spaced out so she had time to think.

The day of the result was the most surreal of her life. In the waiting room, Valentina's husband reminded her that she didn't actually have to get the result. They could continue not knowing. But Valentina had so many symptoms by then that she didn't feel she could put it off any longer. She remembers arriving early and seeing the counsellor walk up the corridor to the clinic room. The counsellor was wearing a Covid facemask and didn't look in their direction. Valentina was desperate to read her expression but couldn't.

Inside the clinic room, the counsellor told them they could take off their masks. When she lowered her own, the counsellor smiled.

'And I knew,' Valentina said. 'When I saw her smile, I knew I didn't have it.'

Valentina's test was negative. She did not have HD.

'It was the most amazing thing. The most extraordinary feeling,' she told me.

All the symptoms that had plagued her and which had accumulated over years were due to other things. Some were pre-menstrual, she thought, but many were due to her anxiety and constant symptom checking. Some melted away with the negative test and some stayed but became less severe.

'I still get flustered in the airport,' she told me. 'I still cannot multitask, but the difference is that it doesn't spiral. When I thought I had HD, I'd walk into the airport and feel my anxiety rising. Then I'd start feeling dizzy and panic. I couldn't walk or think. Now if it happens I don't worry about it, so it just peters out.'

The aftermath of the negative test was both wonderful and terrible. First, Valentina had the amazing moment of telling her children they were all clear. The family celebrated together that evening. But HD is a family diagnosis, so a negative test for one member is inevitably bittersweet. Valentina didn't have it but Evangeline still did. Luca and Camila still might have it. How would she tell Evangeline? They had been comparing symptoms and supporting each other for years. They used to joke about eventually sleeping in twin beds in a nursing home, swearing at each other liberally. Valentina felt guilty because she had escaped and Evangeline hadn't.

The morning after her negative test result, Valentina went to see Evangeline. It was a difficult conversation but Evangeline showed unqualified delight for her sister, as hard as that must have been. The conversation with Luca and Camila was easier. They were overjoyed. Valentina's negative test gave them more hope. They also blamed every bad mood and stumble on HD. If Valentina was wrong, they might be too.

Her father, Philip, was, needless to say, relieved and jubilant for his daughter. Probably the most unexpected outcome of the negative test for Valentina was how her relationship with her father changed. It deepened. For decades, he was the only person in the family who definitely didn't have HD. He was everybody's support. He'd had to watch his wife change irrevocably and, as far as he knew, every one of his children and grandchildren could also have HD. Philip and all the siblings' partners have had to carry a burden. There is no specialised counselling or support for them. Valentina's normal test meant her father suddenly had an ally.

'I can't get rid of him now!' Valentina laughed.

The only person Valentina did not tell was her mother. Vivian had taken Evangeline's positive test badly and Valentina feared her result would only deepen the pain of Evangeline's diagnosis.

'She never asks anything about your test result? Or about the others?' I asked Valentina.

'No. She seems to just assume that the rest of us are clear.'

*

In the 1980s, before the gene for HD was identified and testing became available, surveys of families affected by HD were overwhelmingly in favour of predictive testing.[1] [2] The majority of people at risk said they would take that test as soon as it was offered. In 1983, almost as soon as chromosome 4 was identified as the likely site of the Huntington's gene, the Huntington's Disease Society of America (HDSA) began to create testing guidelines in anticipation that the test would soon be offered. The forty years since has given the HD community plenty of time to consider the value of a predictive diagnosis of an untreatable condition. And despite their original enthusiasm, when faced with the reality of testing, around 90% of the people worldwide who go to a genetics clinic and are offered the test ultimately choose *not* to have it.[3] [4] The uptake of testing among at-risk populations has been estimated to be as low as 5% in France, 9% in Greece, 15% in Australia and 18% in Canada.[5]

There are lots of very good reasons to have a test for a future diagnosis like HD, even if it is untreatable. For couples planning a family, testing provides great relief if the result is negative and if it's positive, they can choose to have PGT. Knowing the potential diagnosis is pending ensures that a person is given the appropriate support and treatment as soon as the symptoms begin, and avoids the risk that early signs might be mistaken for other things, as happened to Valentina's mother. Neurodegenerative conditions like HD eventually rob a person of the ability to make sound decisions. A diagnosis ahead of time offers the opportunity for someone with HD to dictate the terms of their own future care and to plan their financial future, and their family's future, before they are too sick to do that for themselves.

But for all those good reasons to test, most people still don't. Why? To understand this better, I spoke to Dr Shereen Tadros, a clinical geneticist who regularly sees people at risk of HD.

'Often, all I need to do is give people permission *not* to test. They come to clinic thinking it's the responsible thing to do. It gives people peace of mind to learn that they don't have to test,' she told me.

Dr Tadros estimates that more than 80% of people she sees in clinic who have recently learned that they have a parent with HD arrive with a certainty that the right thing to do is to test immediately. After that first consultation, the number who actually take the test plummets to as low as 10%. That experience is the same across most genetics clinics worldwide.

A common reason not to test is the realisation that knowledge cannot be unlearned. There is no turning back from a positive result. The condition is untreatable and a positive result risks a person blaming every future symptom on HD instead of thinking of and looking for more ordinary explanations.

'Sometimes, all a person needs is a physical examination to reassure them that they are still healthy and that's enough,' Dr Tadros said.

It seems as if some people go to the genetics clinic thinking they want a test, but what they really want is reassurance and permission to get on with their lives.

Geneticists are not gatekeepers to genetic tests. Their job is to provide knowledge but also to drill down into the reasons for the test. They want to make sure that when the dark days after a positive test arrive, the person can look back and remember exactly why they wanted to test and how they hoped to benefit from the result.

People at high risk of genetic disease fear for themselves and for their children. There is a sense that knowing for sure will be empowering. But there can be decades between a positive genetic test and the onset of the disease. In theory, those are the years for living, but that underestimates the impact of the diagnosis on the experience of self. Diagnostic labels have the power to create illness in the absence of disease because thoughts, ideas and emotions are enacted through the body. When a person feels an emotion, it is not an ethereal sensation of the soul: it's felt in the body. As goosebumps. As butterflies in the stomach. Confidence, happiness, anger, timidity, upset, confusion – each of these are evident in the person who is experiencing them because our bodies manifest our internal experiences. But that mind–body interaction is not flawless. The advanced warning of an impending disease and the fear this brings can, along with a person's internal image of their brain degenerating, create bodily symptoms even before the degenerative process has begun.

Understanding a process called predictive coding will help explain how fear of illness can be converted into real physical symptoms. This is the process through which our brains make us efficient and safe, but which also, on occasion, causes psychosomatic symptoms. Rather than passively registering the environment around us and absorbing incoming signals like a sponge, our brains use past experience to predict how our bodies will behave in certain circumstances and to interpret sensory signals. All those things we see, hear, feel and smell are quickly compared to mental models or templates stored in the brain, created from learned experience. That comparison allows us to predict their meaning without the need to relearn the rules of the world every single day.

Imagine you are crossing a busy road. The visual information of the car speeding towards us enters along the visual pathways, but at the same time, higher-level processing is very rapidly manipulating that image to assess the type and size of vehicle, its speed and distance. This is predictive coding allowing you to cross the road safely by relying on prior learned experience. Most adults will have crossed enough roads to allow them to make a good estimate of when it is safe, based on their knowledge of what a speeding car looks like and their knowledge of their own walking speed. Of course, not every road crossing is well judged because the brain's inferences are not perfect. Predictive coding is a system of best guesses. It is the reason visual illusions containing two images are seen differently by different people – faced with a puzzle, our individual brains show us the most likely solution based on what we know of the world. That solution will vary from person to person.

Even the healthiest brains in the most intelligent people make regular mistakes and it is out of those that psychosomatic symptoms can occasionally arise. The brain's predictions are not always right and if a wrong prediction is held strongly enough, that can influence how the body feels. Imagine a person has a phobia of needles. They are having a blood test and they are absolutely certain that it will be terribly painful. That can lead to the experience of pain before the needle even pierces the skin. Their strong expectation has overwhelmed their nervous system and produced the pain even without the painful stimulus.

The brain also uses predictions and expectations to be selective about what it pays attention to. While some sensory experiences are heightened, others are filtered out if they are deemed unimportant. This filtering process is the reason we don't usually feel the sensation of our clothes against our skin or hear monotonous background chatter unless prompted to notice it.

Importantly, predictive coding and filtering don't just mould how we experience signals coming from outside, they also alter our experience of internal sensations. Our bodies are awash with white noise. Our beating hearts, inflating lungs, contracting bowels and tingling skin are all available to be felt but we don't routinely pay them much attention. We tune them out because we're used to them. However, a person who has reason to be concerned for their health might start to notice and worry about any one of these internal phenomena. Expectations and attention play a fundamental part in this. Imagine a person who has just learned that there is serious heart disease in the family. Everybody's heart rate rises as they walk up stairs, but having newly learned of this frightening family history this person might be provoked to pay more than usual attention to their own chest. They notice the heart rate rise as they go up the stairs and mistake it for something abnormal. Attention to the body interrupts the normal filtering process to produce symptoms from an otherwise healthy body.

Predictive coding and filtering help to account for how something called the nocebo effect can make people feel physically ill in the absence of disease.[6] Everybody is familiar with the placebo effect – the power of strong belief to alleviate symptoms. The placebo effect has been shown to produce measurable physiological changes, such as lowering blood pressure and slowing the heart. The nocebo effect is the opposite phenomenon. If a person believes a treatment will cause harm, the nocebo effect can produce real physical symptoms out of those expectations. Imagine eating food and then learning it was contaminated in some way. That knowledge is enough to produce feelings of nausea before the effect of any contamination has had time to set in.

These are the biological reasons why a person who has a strong belief that they carry the HD gene could experience the typical symptoms from that belief alone. Valentina described the experience very vividly. She had seen HD in high definition by bearing witness to her mother's slow progression to disability. She knew what symptoms to expect so she looked for them. She stopped tuning out experiences that most of us would dismiss and paid undue attention to ordinary bodily changes. Everybody bumps into furniture from time to time, but for Valentina, every such stumble was a sign. HD has quite an insidious onset and many of the early symptoms are things we all experience, like anxiety, low mood or irritability. Ordinary feelings became very scary very quickly for Valentina.

Valentina was caught in a vicious cycle in which the more she noticed, the more attention she paid. Movement is supposed to be automatic but becomes clumsy if it's watched too closely. That's why we are more likely to fumble at sports if we have an audience. Valentina's body became less efficient the more she paid attention to it. Valentina's mother struggled with the level of organisation required for airport travel and, similarly, Valentina needed only to step inside an airport to feel sick. Her body became harder to read because her stress pathways were activated by anxiety. Every adrenaline-fuelled palpitation, every tremor, every hot sweat took on greater significance in the knowledge of the impending, unconquerable disease.

Valentina was rescued from this trap by a negative test for HD. That was all that was needed to alter her expectations of her body and direct her attention away from her perceived symptoms, so that they immediately lessened. She still felt flustered at the mere sight of an airport but knowing she didn't have HD prevented it from escalating out of control.

But what if she had tested positive? She would likely have gone on assuming that all her symptoms were due to HD. She may have become more symptomatic instead of less. The fear of HD would surely have ruined her life, many years before the actual neurodegeneration set in.

'There's a big difference between something being at the back of your mind and at the forefront,' Dr Tadros reminded me. 'After a positive test, every time you drop a cup, every time you're grumpy with your kids, or forget something, or double-book yourself, you switch from thinking *has it started?* to *this is it, it's started.* After a positive test, people are at risk of losing many healthy years and lots of fun just through being convinced it's started when it hasn't. People need a glimmer of hope. It keeps them going.'

*

Emily was only twenty-six when she tested positive for the HD gene. She knew exactly why she wanted to be tested and doesn't regret her decision. That is not to say that any of it was easy.

Just as it did for Valentina, the discovery that HD was in Emily's family came as a surprise. Her great-grandmother was believed to have dementia and Parkinson's disease. But somebody must have doubted those diagnoses because, when she died in 2002, a post-mortem was carried out and her real diagnosis, HD, was uncovered. When that discovery was made, Emily was five, her sister was two, their mother was in her thirties and their grandmother was in her fifties. Three generations of the family had been born without anyone knowing they were at risk of HD.

Emily believes that her grandmother was already symptomatic when she learned of her mother's diagnosis. She behaved strangely. She was fixated on certain things. She wore the same coat every day. Even if she was sweating profusely, she wouldn't take it off. But her formal diagnosis only came when she developed speech and walking difficulties and the typical jerky choreiform movements.

Emily's mother's decline was different. It started in her early forties and manifested as personality change, psychiatric and cognitive symptoms. She became abusive and physically aggressive towards Emily and her sister. She was disinhibited in the street. The motor symptoms came later. These days, she struggles to swallow or speak. Her walking has declined a great deal and she is dependent on her family.

Emily's parents started speaking to her about HD from a young age. She's grateful for that. As soon as she turned eighteen, which is the minimum age at which an asymptomatic person can have a predictive genetic test for HD in the UK, she made an appointment with the genetic counsellor. They decided together that she wasn't ready to test. Emily's relieved by that now. Watching her grandmother and mother decline was a lot for a teenager to handle without the additional trauma of a positive test.

Emily put off testing but over the years that followed made occasional appointments with the counsellor to explore the possibility again. Ultimately, she decided to wait until her grandmother had died before she tested. After her grandmother had passed away, Emily gave herself time to grieve, worked with her genetic counsellor and tested just over a year later.

Emily isn't one of the people who could be sustained by hope. She was never able to turn off thoughts of HD. She thought daily about getting sick. She was absolutely certain she had the gene. She wanted to be prepared and could only do that by testing. But her biggest reason for testing was to be certain that she and her sister would be the last people in her family who had to face this dilemma.

'I always said to my mum, if I have the power to stop this disease with me, that's what I will do,' Emily told me.

Emily didn't cry when they told her the test was positive, but she did cry later at the thought of having to tell her mum. It was as bad as she'd expected. Her mother screamed and nearly collapsed when she learned she had passed the disease gene to her daughter. Emily took it harder than she expected too. She had planned to continue working but found she was just too upset, so took time off. She didn't know how to feel. It was a confusing time. Other people's responses made her feel like she'd died. They sent her flowers, were sympathetic for a few weeks, like a funeral period, and then they forgot.

Emily's family were their own support.

'My dad's an incredible man,' Emily said. 'He was strong when I told him my test result. He was there for Mum and me.'

Emily's father was once a nurse and now cares for his wife as her HD progresses. They are a close family. Emily and her father do advocacy work and they both spoke of their experience of HD at a charity event for young people at risk. Emily's father spoke at length about his wife's diagnosis but didn't mention Emily. That upset Emily until her mum told her that, behind closed doors, her father had cried and cried and cried for her.

'Has knowing you have the gene changed how you live your life?' I asked.

'It ebbs and flows,' she said. 'Sometimes I think this was supposed to be and that it will inspire me. It has definitely made trivial things less important and helped me live my life for now. But I also get moments when I'm lying in bed and realise that some day I won't be able to get out of bed. Some day I will rely on others for everything.'

Emily is due to get married soon. She met her fiancé when she was twenty, before she tested. She told him about her family history three weeks into the relationship and he was not the least put off. Many of her friends in the HD community avoid relationships because they are so afraid of the consequences of telling someone. Emily advocates for being honest and vulnerable in relationships and it has worked well for her.

Emily and her partner plan to start a family in a year or two using PGT. She sometimes worries about the morality of her decision to have a child. Anonymous people on social media, commenting on her HD posts, have criticised her for planning to have children. She wonders herself if it's right to have a child who will have to watch their mother suffer in the way she has seen her own mother suffer. Then she remembers her own life.

'If my mother had thought that, I wouldn't be around and I've lived a great life,' she said. 'I've done a lot of great things. My life has been worth living.'

During the pandemic, Emily's grandmother was confined to her nursing home. Lockdown meant the family could only wave at her through a fence. Emily couldn't bear it, so she took a job working weekends in the nursing home alongside her full-time job. The work was hard and most of it was not spent with her grandmother but with other residents of the nursing home. But, for Emily, it was well worth it for the lunch breaks they had together.

Emily has seen the worst of what HD has to offer. She saw her grandmother bed-bound. She saw her mother's personality change and her physical abilities decline. But she hasn't let it overshadow everything.

'Life doesn't stop with a positive HD result and life doesn't stop with HD,' she told me. 'If I am going to stare at the ceiling for years I want to have really good memories to relive.'

Emily is hoping for a medical breakthrough and there is plenty of time for it to happen. But, just in case, she intends to make the richest memories she can. In response to learning her test was positive, she climbed Mount Kilimanjaro.

★

A predictive diagnosis is not actually a diagnosis. It is a warning about a future diagnosis, but one that potentially has all the impact of the disease itself. Its strength is that it gives people a chance to plan and removes uncertainty. But it also means that someone is forced to confront the implications of a disease that might be decades away. To be forewarned may be to be forearmed – but it may also be a burden.

HD testing is an example of the gold standard of predictive diagnosis. Worldwide, charities supporting HD families are unequivocally in favour of making testing easily available to those at risk, but they are also clear that it should only take place in a well-managed way.[7] Testing is never urgent. Genetic counsellors can spot the vulnerable people in need of more time or more support. They can slow the process for those who don't seem ready.

The repercussions of a positive test are more profound and wide-reaching than many people predict. Understanding that is a fundamental part of the counselling process. There are obvious psychological consequences for some. Having learned of a pending diagnosis, a person might live the rest of their life as if the disease has already started. Depression and suicidal ideation are not unusual in those who test positive, but also in those just facing testing.[8] Valentina told me that she contemplated contacting an assisted dying facility if her test had come back positive. I hope and believe she would not have gone through with it, but it is not an unusual response. A positive predictive diagnosis also runs the risk that every future experience, even the commonplace, is mistaken for an inescapable part of an incurable neurodegenerative condition.

But the implications go beyond the psychological. Driving, insurance, employment can all be affected. Most countries require regular reviews of a person's fitness to drive and, once symptoms are deemed severe enough, they are prohibited from driving. In the UK, people taking out high-value insurance have to declare their positive test. In the US, it is recommended that a person considering being tested obtains life insurance, disability insurance, and long-term care insurance prior to testing, as a positive result will impact their ability to take out a policy in the future. And while there is no requirement for a person to tell their employer they have tested positive for HD, geneticists still have to consider the responsibility a person holds in their job, especially those who are in high-risk roles (doctors, for example, or those working with dangerous machinery). If the geneticist feels a person is a risk to others because of their HD symptoms, they could potentially break confidentiality and reveal the test result to third parties. Confidentiality is built into genetic testing, but it isn't guaranteed.

In 2019 a woman sued an NHS trust that had not informed her of her father's HD diagnosis. Doctors had asked for permission to share the diagnosis with his daughter, pregnant at the time, but he had refused. The woman, who later learned she also carried the HD gene, argued she had a right to know and that she would have terminated her pregnancy if she knew she was at risk.[9] She didn't win the case. A doctor's duty of care is to their patient, not to their patient's family. However, the judge reinforced the need for geneticists to undertake a balancing exercise to weigh up contradictory duties of care. Which means that while most results will be kept private, if their patient's disease makes them a risk to the public or to another individual, a healthcare professional may be able to make a strong enough argument to override the patient's right to confidentiality and release the result of a test. In other words, a person does not have an absolute right to veto disclosure of their own genetic results.

So, genetic test results can place healthcare professionals in a morally difficult position. Even when the law prevents such disclosure, doctors can feel compromised if they have knowledge that could be helpful to others but which they are forced to withhold. There is also the question of which other interested parties might seek to gain access to the results of genetic tests. In the US, there have already been cases of law enforcement agencies demanding results.[10] It is also worth knowing the rules where you live because they differ between countries. In New Zealand, insurers can use genetic results to legally discriminate against applicants.[11]

Clinical geneticists and the HD community have been extremely thoughtful in the development and control of the management of predictive genetic testing for the people they represent. They require a high standard that will act as a good template for other communities in their position. The system is careful, slow and considered. But it is unlikely that level of care can be sustained as a great deal more tests become available for many more conditions. HD is a predictable disease and, as a burden to health services, the numbers affected are small in population terms. In Australia, North America and Europe fewer than 8 in 100,000 people have HD. In Asia and Africa, that figure is fewer than 1 in 100,000.

But how will the delicate art of managing a predictive diagnosis work when applied to much larger groups of people and those with much more unpredictable diseases? What of common medical disorders for which genetic testing is now available, like Alzheimer's dementia? Carriers of the APOE e4 gene have a substantially increased risk of Alzheimer's disease. Crucially, as things stand, there is no cure for Alzheimer's and the current treatments make little difference to how the disease progresses. Like with the HD gene, people who learn they are APOE e4 carriers face an incurable neurodegenerative disease at an unspecified point in their future. But with an important difference: everybody with the HD gene *will* develop HD if they live long enough, but APOE e4 is only a risk factor for Alzheimer's. Not every single person who carries the gene will get the disease. Some people who test positive for the dementia gene might live weighed down by the thought of impending dementia entirely unnecessarily. Most public health services don't test asymptomatic people for APOE e4 precisely because of all the uncertainties – but the test can already be purchased in the private sector and in over-the-counter genetic tests with no specialist counselling required.

A genetic test result may belong to an individual but it impacts the whole family. It has implications for those who didn't consent to the test and who may have preferred not to know. A person testing positive immediately moves their own children into the high-risk category for testing positive too, and the children have no protection against that knowledge. That's a large amount of hope robbed by a test that person did not choose to have. It is even more complicated if an adult child wants to test but their parent hasn't. If that child tests positive they will reveal that their parent must be positive too. And there are no privacy laws to prevent one family member revealing their own diagnosis and, in doing so, exposing the whole family.

Then there are consequences of a positive test that are hard to anticipate because not all the implications of predictive testing have been fully challenged yet. At what stage should a potential future spouse or partner be advised of a genetic risk? They surely have a vested interest if their child might also be affected. If the partner has not been informed and has an affected child, might they not have strong grounds for complaint and even legal action? If the spouse has been informed, could they use it to gain custody of their children in the event of the breakup of the relationship? Numerous parties have an interest in the result of a genetic test. Family, partners, future partners, children, future children, employers, insurers, banks, mortgage lenders, educational loan companies. For now at least, their interests are being held at bay in favour of the right to privacy of the patient. Will it stay that way?

There will also likely be covert discrimination against people who test positive. Predictive testing is so new that even doctors don't quite know how to respond to a positive test. HD testing may have been around for decades, but it's a rare disease. It is only now that genetic testing has become more widely available for a huge range of conditions (like Stephanie's KCNA1 variant) that reacting to predictive tests has become part of every doctor's job.

'Having the gene feels like a black mark on your medical records,' Emily told me. Since Emily learned she carries the HD gene, she has struggled to get routine medical care from doctors who appeared to be scared to treat her. Healthcare workers haven't always understood the difference between a positive predictive test for a disease and actually having the disease. Emily was denied treatment for ADHD on the basis of her positive test. Her predictive diagnosis seems to intimidate people who should know better. It bodes poorly for how people with a positive predictive diagnosis will be received in the wider world if medical care is unduly influenced by and scared of it.

A predictive diagnosis of HD, dementia and other untreatable genetic disorders has value, of course. If only to allow people to plan their future. But the potential negative consequences should not be underestimated. Knowing you are at high risk of a disease could change how you use your body and how much you trust it. Worry and uncertainty creates fertile ground for the misinterpretation of every normal illness and bodily change. A medical label is not an inert thing. It has the power to make people ill even when the body is otherwise healthy.

Illness, pain and death are inevitable for everyone, but we don't live our lives thinking about them. If we did, it would probably paralyse us. As we develop techniques to do more predictive tests, we must keep in mind that the fear created by the knowledge of an imminent disease could change the trajectory of a person's life, stealing their ambitions. The threat of a future diagnosis risks casting a shadow over every otherwise carefree day, as it did for Valentina until she was freed by her negative test. The tenet of medicine is to do no harm. The psychological and practical consequences that go along with predictive diagnosis do not always fit well with that tenet.

Suggestions that predictive genetic testing should be dialled back or withheld until more is known and better systems are in place to support people can be met with accusations of paternalism. The argument is that people have the right to know more about their future health, if it can be known. But in our excitement to make the most of this new scientific capability and to allow people every choice, I cannot help but feel that a large part of the HD story isn't being heard. I was very struck when Dr Tadros told me that, often, all she has to do is give people permission *not* to test. Many people's first instinct is to assume that knowing is empowering and testing is the responsible thing to do, but one frank conversation with a genetic counsellor reveals that a large percentage of people do not want to risk a positive test hanging over them. They choose hope. Before testing is available to a great deal more people for many more conditions, there is a pressing need to find a better balance between protecting autonomy and protecting a person's right *not* to know.

Much is rightly said about the struggle of living with undiagnosed symptoms. A diagnosis that explains symptoms usually comes as a relief. But that is not what predictive diagnosis does. It warns you about future symptoms. It doesn't tell you when they will happen or how they will start, just that they probably will. Once you know those symptoms are on their way, you can't un-know it. Before a diagnosis is confirmed, any future is still possible. After it is confirmed, the future is limited. Knowing for sure may prove to be more painful than uncertainty.

2

Lyme Disease and Long Covid

Lyme, Connecticut, is a beautiful, leafy, rural village by the sea. Artist Polly Murray was in her twenties when she and her husband Gil first relocated there in the late 1950s. Miles of woods surrounded the large white house where the Murrays raised their four children. The family loved that house and those woods. They loved the outdoors. The children played in the kiddie pool on the lawn. They tobogganed down hills in cardboard boxes. As they got older, they built forts in the woods and, as a family, they went on walks through the trees and long grass, giving names to all their favourite spots, like 'Deer Patch' and 'Indian Lookout'. In the winter, the children would ice-skate on nearby lakes. In the summer, they picnicked on the beach and swum in Long Island Sound.

In many ways, a life in Lyme was idyllic for the Murrays, but sadly that wasn't the full story. Almost as soon as Polly had moved to Connecticut, she started to get sick. She was plagued by mystery illnesses: flu-like symptoms, rashes, aches and pains. The children's health was not much better. Headaches, fevers, joint pains, sore throats, conjunctivitis, swollen glands, gastroenteritis, swollen fingers, lethargy, muscle aches. No sooner had one symptom disappeared than another emerged, or one member of the family recovered and another fell sick.

Naturally, Polly sought the advice of local doctors for all these complaints but she didn't get much satisfaction from those visits. Various diagnoses were offered, along with occasional doses of penicillin, but nothing really helped. As the years passed, the illnesses continued. Polly started to sense her doctor sigh every time she turned up in his office. Polly was convinced their symptoms were connected by a single diagnosis but nobody else seemed to be terribly interested in why one family should be so unlucky. Even Gil was of little help, preferring to turn his back on these nagging concerns.

Polly was alone in her search for a diagnosis but she was not unarmed. Highly intelligent and with a good grasp of science, she channelled her frustration into research, making shortlists of possible diagnoses. Those she brought to doctors throughout New Haven and, later, Boston. She wondered if she could have lupus. Or a virus? Or a streptococcal infection? One New Haven doctor seemed to go along with her suggestion of lupus – an autoimmune condition – but after that was dismissed the doctor didn't address her other theories. Another doctor mooted vitamin C deficiency and offered supplements which Polly dutifully took, to no good effect.

One of Polly's theories was that she might have something akin to Rocky Mountain fever, a tick-borne disease. Her problem seemed very similar and the area surrounding Lyme was thick with ticks. They lived on deer, of which there were many in the woods in the region. She dug the ticks from the children's skin on a regular basis. Engorged ticks fell from the fur of family pets. But she could not seem to interest her doctors in this comparison.

Over three decades, the family were thoroughly investigated but tests always came to nothing. The children's joints swelled. Sometimes they needed crutches to help them walk. All the family got skin, eye and throat infections. In 1971 alone, seventeen years into her mystery illness, Polly was hospitalised three times. During one hospital stay, a doctor asked her why she always looked so serious. With her chronic ailments, Polly told him, she didn't feel there was much to smile about.

Polly knew the doctors didn't like it when she did her own research. One of them called her a doctor-chaser. Another commented, 'I suppose you think this is some new disease. Why, they might even call it Murray's disease!'[1] Polly did not feel able to defend herself but cried later, when she was alone.

One of the Boston doctors asked Polly if she had considered that the problem might be psychosomatic.

'You know, Mrs Murray, sometimes people subconsciously want to be sick,' he said.[2]

It certainly didn't feel that way to Polly but when he suggested she see a psychiatrist, she agreed. She always did what the doctors asked, even when it seemed counterintuitive. She didn't care *what* was wrong with her, she just needed a 'known entity' so she had a sense of what she was up against. She accepted a three-week admission to explore psychological treatment. Actually, it was not entirely a waste of time. She got a small amount of relief from some of the conversations she had with one of the more insightful psychiatrists in the unit. Not that it made her physical symptoms any better. Instead, it taught her that she should take her medical research underground. After leaving hospital, she walked into the Yale medical library and pretended she belonged there.

And there was more than just the Murray family's illnesses for Polly to investigate. Over time, she had become aware that dozens of other people in her area had equally unexplained ailments. A neighbour had been hospitalised multiple times for unexplained symptoms. Another had been plagued with skin rashes for years. Multiple children had swollen joints.

Even local pets were sick. Polly kept careful records of all these symptoms. She was a quiet woman, but she was also dogged and meticulous. Polly was convinced that there was a new disease in Lyme.

Polly had first fallen ill in 1954. It was not until twenty-one years into her unofficial investigation that she found some allies. Finally, in 1975, she received a more serious response to her concerns from the Connecticut State Health Department. Somebody else in the community had reported similar suspicions and a doctor, Allen Steere, was appointed to investigate the phenomenon. It didn't take long for him to see what Polly saw. In particular, he noted an exceptionally high rate of juvenile arthritis among the children of Lyme. He agreed there was a novel illness in rural Connecticut and that it was spreading like an infectious disease.

It took seven more years for researchers to confirm the cause. In 1982, the bacterium responsible, *Borrelia burgdorferi*, was isolated and named after Dr Willy Burgdorfer, the scientist who first discovered it. Burgdorfer found the bug in the midguts of deer ticks and it was later identified in the blood of people who were sick. Nobody actually suggested calling the medical disorder caused by *B. burgdorferi* 'Murray's disease' in honour of Polly. They stuck with an old tradition and called it after the town where it was first described: Lyme.

*

Lyme disease is an infectious illness caused by a spiral-shaped (spirochete) bacterium. It's transmitted to humans by a tick bite, particularly from the blacklegged tick that often live on deer. The typical clinical presentation is a characteristic bull's-eye and spreading rash that appears at the site of a tick bite days or weeks later. Ticks have a fondness for damp crevices – such as the back of the knee, the groin, the armpits and under the breasts – so these are the places where the rash is most likely to appear. Flu-like symptoms, muscle aches, chills, headaches and tiredness can appear soon after the rash. For some, that is the entirety of the illness, but for others the disease evolves into a multi-system disorder. Swollen arthritic joints are common in more severe disease, as is facial palsy. Nerve root involvement can cause crippling pain akin to sciatica. Inflammation of the brain and spinal cord can be life-threatening and lead to long-term disability. If the infection extends to the heart, it can cause palpitations, dizziness and shortness of breath. The true incidence of Lyme disease is unknown in the UK because it is not a notifiable disease, meaning there is no requirement for doctors or laboratories to report newly diagnosed cases to any overseeing body. However, data from positive tests is routinely collected and from that it is estimated that there are 2,000–3,000 new cases diagnosed yearly.[3] Lyme disease is notifiable in the US, where there are approximately 60,000 new cases of lab-proven Lyme disease per year.[4]

As an infectious illness, one might expect that Lyme disease should be a fairly straightforward diagnosis. Surely a bacterial infection is something concrete? It would seem reasonable to assume that a person shown to be infected by the bacteria could be diagnosed with Lyme disease and a person in whom the bacteria cannot be detected should be assumed not to have it. But, in fact, the question of who has and who does not have Lyme disease has been a matter of great controversy. So much so that in 2007, the *British Medical Journal* published an article referring to the debate over Lyme disease diagnosis as the 'Lyme Wars'.[5] It described an acrimonious standoff between two sets of physicians. One group argued that Lyme disease was tragically underdiagnosed and another that it was overdiagnosed. Nearly twenty years after that article was published, that argument rages on.

Polly Murray's battle to get a diagnosis was just the very beginning of the fight over Lyme disease.

Key to understanding the Lyme wars is to know that almost all diagnosis is, at least in part, subjective, which makes it open to uncertainties, mistakes and exploitation. No matter how many sophisticated tests are involved, diagnosis remains a clinical art that is inherently intuitive and heavily reliant on a doctor's interpretation of the patient's story and examination. We are in an era of burgeoning technology in which tests are perceived to add accuracy to diagnosis and, while that can be the case, they can also do the exact opposite. Tests can pretend certainty while actually *contributing* to error. Lyme disease is only one example of how an apparently objective diagnosis can incite entirely opposing views because of the subjectivity of diagnosis and questions about the reliability of test results.

Lyme disease was identified more than forty years ago so there has been a lot of time to perfect a diagnostic test. The bacteria *B. burgdorferi* don't grow easily in blood culture like other bacteria and are often present in scanty amounts in the body of someone with Lyme disease, so they're hard to find. For that reason, the focus of testing is to look for a particular antibody that indicates the immune system has mounted a defence against the bacteria, rather than looking for the bacteria itself.

Both the Centres for Disease Control and Prevention in the US (the CDC) and the National Institute for Health and Care Excellence (NICE) in the UK have guidelines for optimum testing, which takes place over two stages.[6 7] This means that a single blood sample given to investigate the possibility of Lyme disease will potentially undergo two different types of testing. The first test is called enzyme-linked immunosorbent assay (ELISA) and the second is the Western blot. Both look for antibodies directed against surface proteins (antigens) on *B. burgdorferi*. The ELISA looks for a very wide range of possible antibodies directed against lots of different antigens. If the ELISA is negative that is interpreted to mean that Lyme disease is very unlikely, so no further testing is needed. (There are multiple caveats to this that will be discussed.) However, if the test is positive that does not automatically mean that a person has Lyme disease. The antibodies it detects are also produced in response to other infections and immune disorders, so a positive ELISA could occur for reasons other than Lyme disease. These are false positives.

The Western blot test is used to look for a much smaller range of antibodies that are much more specific to *B. burgdorferi* infection. This second stage of testing is only required if the ELISA is positive. The purpose of it is to weed out any possible false positives found on ELISA. To be considered to have tested positive for *B. burgdorferi* consistent with potential Lyme disease, a person must have a positive ELISA followed by a positive Western blot.

So why, if there is two-stage testing, refined over years and ratified by the CDC and NICE, is there so much controversy over the diagnosis? One would not expect a test to be right all of the time, but surely it can be trusted to be right most of the time at least? The heart of the issue is that the ELISA and Western blot are not *diagnostic* tests, even if they might appear that way. They are only a piece of a diagnostic jigsaw and are meaningless without the much bigger part of that puzzle – the clinical story. This is the case for most diagnoses, not just Lyme disease.

Tests, no matter how sophisticated, are not usually used to make a diagnosis but rather to provide supportive evidence for a clinical theory. Results are interpreted differently according to the pretest probability that the person being tested has the disease. Pretest probability will be important to later diagnoses in this book too – it essentially refers to the likelihood that a person has the diagnosis under investigation, as determined by their specific symptoms and full circumstances. A person has a high pretest probability of having Lyme disease if they have the typical symptoms and have been in a Lyme disease endemic region. They have a low pretest probability if their symptoms are vague and there was no definite exposure to Lyme disease. A positive test is usually only treated as a true positive if the pretest probability that the patient has the disease is high, as judged by their medical history. The CDC and NICE do not only say which tests should be done to help confirm the diagnosis, but also that the results must be interpreted in the context of the patient's full clinical history.

The doctor's understanding and interpretation of their patient's story is the first stage at which subjectivity enters the diagnostic process. The second stage is when tests are processed by the lab and then interpreted, initially by a lab doctor and later by the diagnosing doctor. You might think that medical investigations are there to prove a diagnosis and to overcome a doctor's uncertainty or bias. That is the case but only to a point. Because, despite the appearance of objectivity, medical investigations, the ELISA and the Western blot included, are rarely as standardised or straightforward to interpret as you may think,[8] and few give a yes/no answer. Rather, they present the possibility of a diagnosis, with multiple provisos. Many test results are as ambiguous as clinical assessments.

Medical investigations are confounded by many variables. In the case of blood tests, for example, ethnicity, diet, exercise, alcohol, hydration level, medications and other diseases can all affect test results. Equipment and analytical processes vary between labs, sometimes even leading to different results for the same test in the same patient. Results are usually presented as a set of values called a normal range, meaning two people can have entirely different measurements and still both be considered normal. It is ultimately up to the doctor's clinical experience and good judgement to know all the confounding variables for the test they ordered and to keep them in mind when making a diagnosis. It is up to the lab to optimise how they do the tests to produce the most reliable results.

The tests for Lyme disease speak to this challenge. There are lots of ways for tests to produce misleading results, even when carried out within the guidelines. A lab that sets its test to look for antibodies to a very large number of non-specific antigens, many of which are also seen on other microorganisms, will produce a very sensitive result with lots of positives, many of which are false positives. If a person is very sick either due to another infection or any autoimmune disorder that creates high volumes of circulating antibodies, that can easily cross-react with the test and create a false positive result. False negatives can occur if the test is calibrated to look for antibodies to too few antigens. Different strains of *B. burgdorferi* live in different geographical locations, so labs that test for the wrong strain could miss them. If the test is done too early, it may produce a false negative result by pre-empting the immune response. The very latest testing techniques have reduced some of the effect of the confounding variables described here, but haven't eradicated them.

Another key point is that a blood test for *B. burgdorferi* is *not* a test for Lyme disease. It only indicates that at *some point* in a person's life they have been exposed to the bacteria. They may never have developed the actual disease. The New Forest in the UK has a high level of *B. burgdorferi*-carrying deer ticks. A study of forestry workers in the New Forest showed that 25% of them had positive two-stage tests for the bacterium but most reported no symptoms of Lyme disease.[9] These people had worked in a *B. burgdorferi*-rich area for many years so had likely been repeatedly exposed and developed immunity without contracting the disease. Antibodies to past infection or simply exposure do not necessarily disappear. We retain immunity – that's why vaccinations work. All an antibody really says is that there has been exposure to an infectious agent in the past and exposure alone doesn't mean that the bacteria ever caused sickness.

So a positive test that detects an immune reaction does not make a diagnosis and a negative result does not necessarily rule it out. As impressive as our many medical investigations and technological capabilities might look, more often than not, diagnosis is a clinical call. A proven tick bite isn't essential and neither is a positive test, but typical symptoms and known exposure are.

*

Sian lives in Wrexham, a city in Wales. She has spent many years severely disabled by something called chronic Lyme disease (CLD). This is said to be a subset of Lyme disease that causes long-term symptoms from which it is hard to recover. (There is controversy about this definition which I will come to shortly.)

'I don't think anyone in Wrexham had ever been diagnosed with Lyme before me,' she told me. 'I think that's why my doctor didn't think of it.'

Sian was once above averagely fit. She was a physical training instructor in the army who took part in triathlons in her spare time. So it was unexpected when, in 2014, her health changed abruptly. On day two of a city break with her husband, Sian woke with flu-like symptoms, fevers, chills and shakes. She was so sick that she had to stay in bed for the entire week of the holiday. On returning home, Sian consulted her GP who suggested she had the flu and advised her to rest. She did so, but instead of getting better, she got worse. She developed brain fog. She couldn't think, speak or formulate her thoughts. Her heart started to race. Her muscles were weak. Her body tingled. It evolved quickly. Symptoms spread and multiplied.

'I was told I was having panic attacks,' Sian said. 'I worked out that I had about ninety to a hundred different symptoms. I felt like I had ants running up and down my legs. I was jumbling my words. It was as if I'd been poisoned. I could barely take the kids to school. I couldn't even walk the dog.'

Sian consulted her doctor regularly as her symptoms accumulated. Various diagnoses were considered and then dismissed. Tests were all completely normal, leaving Sian frustrated.

'How could I feel so bad and nothing show up?' she said.

Her doctor wondered if she was depressed. Her family briefly wondered that too.

'Could you relate to the possibility of depression?' I asked. As a neurologist, I meet a great many people with rapidly escalating physical symptoms that have a purely psychological cause and much of what Sian was telling me sounded familiar.

'No . . .' she hesitated. 'I *had* been depressed. A long time before. There was a house fire. We lost everything. My husband was overseas so it was just me, the kids and the dog, but we got out. We were all fine. After that, I did have PTSD. I couldn't sleep. But I saw a counsellor and that was the end of that.'

The fire happened two years before the onset of Sian's illness.

'Did the illness in 2014 feel different to PTSD?' I asked.

'It was very different. I wasn't depressed, I was *distressed*. There was something wrong with me and the doctors couldn't find the cause.'

At one point, Sian was diagnosed with chronic fatigue syndrome and was told doing more tests would be a waste of time. That was when she turned to the private sector for help.

She saw a thyroid specialist, hormone specialists and infectious disease specialists. She was not insured so it was all at her own personal cost. She spent thousands of pounds on more tests that came back all clear. Despite the lack of answers, the pursuit of a diagnosis provided some relief because it gave her focus and hope. By then, Sian was housebound. Her army husband was stationed overseas so she relied on her family to help with her three children, then all aged under ten.

It took two years for Lyme disease to enter the frame as a possible diagnosis for Sian. She didn't live in a Lyme endemic area and had not had the characteristic bull's-eye rash. She had heard of Lyme disease but since the official descriptions of it only list a handful of symptoms, compared to her hundreds, she had dismissed it. Then a family member directed her to a television programme about the disorder and, watching it, Sian saw her illness reflected back at her. There were pictures of ticks. For the first time, Sian recalled seeing a small black insect in the shower two days before her holiday. She knew immediately that insect was a tick and that it must have bitten her.

'I was a hundred per cent certain I had Lyme disease when I saw that programme,' Sian told me.

She consulted her local NHS doctor the next day, telling him her theory. The doctor agreed it was possible. He treated her with two weeks of antibiotics. It didn't help. The doctor took that to mean she didn't have Lyme disease but Sian didn't see it that way. She undertook research and decided she just hadn't had enough antibiotics. At her request, her doctor gave her a second round, but they didn't help either. Still certain she was right, Sian contacted a Lyme disease charity for their advice. They advised her she needed intravenous antibiotics. She went to her doctor again. This time he did a blood test for *B. burgdorferi*. The test was negative. He didn't want to give her intravenous antibiotics, but Sian was sure she needed them and set about proving she had Lyme disease, whatever the cost.

Sian described all of this to me nine years after she first became unwell. She was still housebound, although her brain fog had cleared considerably, allowing her to give a fluent account of her illness, which would have been completely impossible at its peak. I asked Sian, if the tests were negative, if she hadn't seen the tick bite or the rash, didn't live in an area known for *B. burgdorferi*-harbouring ticks and the treatment didn't help, why she was still so convinced that she had Lyme disease?

When Sian's local doctor failed to prove the diagnosis, following advice from the Lyme disease charity she had spoken to, she asked a friend to take her blood and sent the samples to two labs in Europe and one in the US. She paid in excess of £3,000 for each of the three sets of tests. All three labs returned positive tests for *B. burgdorferi*. It stood to reason for Sian that these three positive tests overruled her negative test in the UK. She took the results to her GP and told him she planned to have more antibiotics. But because two courses of antibiotics had already failed, her GP refused to supply a third. So Sian approached a private clinic who did prescribe them. She immediately felt better and was able to take the dog for a walk for the first time in two years.

But within forty-eight hours, she was worse than ever.

'It was the Jarisch-Herxheimer reaction,' Sian told me.

When a spirochete infection like *B. burgdorferi* is treated with antibiotics, the death of the bacteria can trigger an immune response in some, called the Jarisch-Herxheimer reaction, which manifests as a flu-like illness. It's a scientifically acknowledged sign that indicates that the antibiotics are working. The deterioration is transient, lasting about twenty-four hours, after which there should be significant improvement as the bacteria finally die. But Sian didn't recover, which is not what one would expect if her abrupt deterioration was indeed due to the Jarisch-Herxheimer reaction. Still, she took it as further proof she had Lyme disease, so she remained on a cocktail of antibiotics for a further six months without getting better. More blood was taken and sent to the US at great financial cost to Sian. Further positive results for *B. burgdorferi* and a host of other infections prompted the private doctors to suggest yet more antibiotics. Her health went up and down. She had better days but she never had good days.

Sian never lost the conviction she had Lyme disease. When her UK-based private doctor reached the limit of antibiotics they would give, Sian decided to seek help in the US, where Lyme disease is much more prevalent. She flew to Washington DC to meet Dr Joseph Jemsek, a US-based specialist in Lyme disease, known for developing a treatment regimen called the Jemsek protocol. He reviewed Sian's story and agreed that she had Lyme disease. He duly started her on the protocol – a pulsed regimen of antibiotics. Some days, she took antibiotics, and on other days, antipyretic drugs that reduce fever, like ibuprofen or aspirin. As Sian explained it to me, these were the same antibiotics she had already taken but, as she understood it, the pulsed treatment programme tricked the bacteria out of hiding.

Sian has been on the Jemsek protocol for six years now. There is a pattern to how she feels. While taking the antibiotics she feels terrible and when she stops them she feels better. But if she stops them for too long, she feels bad again. She believes she would die without them.

'When I come off the antibiotics, that's when I feel the benefit of them,' Sian told me.

'How much better do you get?' I asked.

'Eighty per cent better.'

'So you can leave the house?'

'No. I still can't leave the house.'

Sian has concerns about the testing and treatment of Lyme disease in the UK. She doesn't trust the accuracy of the laboratory testing and fears that doctors are too constrained by regulations to provide people with the antibiotics they need. Many people with chronic Lyme disease and many Lyme disease charities worldwide agree with her. Infectious disease doctors in the UK have told Sian that she does not have Lyme disease and have criticised the treatment she is receiving. But Sian only really trusts her US specialist, Dr Jemsek.

Joseph Jemsek is a controversial figure. He is loved by many Lyme disease patients but also has detractors. Complaints from patients have brought him very close to losing his medical licence. In 2006, the North Carolina Medical Board charged Jemsek with professional misconduct for the improper diagnosis of Lyme disease and improper long-term use of antibiotics. The medical board threatened to suspend his medical licence but said they would put that on hold if he agreed to alter his practice to meet certain conditions.[10] Jemsek complied with the board's order and has since relocated to Washington DC.

Sian speaks to Jemsek by phone every few months and travels to the US to see him once a year, pandemic allowing. When they speak, he listens carefully and encourages her to list all her symptoms every time. She could not say enough good things about him to me but also spontaneously raised the subject of his misconduct ruling. I asked her what she thought about that.

'They punished him because he was helping Lyme disease patients,' she said.

'But why would somebody be punished for helping patients?' I asked.

'They think he's treating people for a condition they say doesn't exist.'

Sian is right in some respects. Most mainstream doctors do not think that the subcategory of Lyme disease called 'chronic

Lyme disease' has anything to do with a *B. burgdorferi* infection. But she was not right about the reason for Dr Jemsek's threatened suspension. Actually, that came about after several patients suffered harm as a result of his unconventional use of antibiotics. He used them off-licence, meaning he gave them to patients in ways that were not officially sanctioned because they had never been tested for efficacy or safety in clinical trials. At the state medical board hearing, several patients testified that Jemsek's aggressive, long-term use of powerful intravenous antibiotics almost killed them. One man said that he thought Jemsek's negligence resulted in his wife's death. One thirty-year-old woman was said to have been misdiagnosed with Lyme disease by Dr Jemsek and subsequently spent four weeks in an intensive care unit after acquiring a multi-drug-resistant bacteria caused by overuse of antibiotics.[11]

Sian feels certain she is getting better. She benefits from talking to Dr Jemsek and from the community of patients she has met through him. She continues to send blood to the private lab in the US to recheck her Lyme status on a regular basis. She has not yet had a result come back completely clear of all infections but, now that she's feeling better, she plans to send another sample to the US soon and hopes that will be the one that gives her the all-clear.

'All I want is to take the dog for a walk. A nice long walk. And I want to go out with my friends instead of talking to them on the phone,' Sian told me. It didn't seem too much to ask.

★

Lyme disease is thought to have a misdiagnosis rate of 85%. In 2019, an observational study carried out at the Johns Hopkins School of Medicine looked at 1,261 people referred with a diagnosis of Lyme disease and found that there was no evidence of active or recent *B. burgdorferi* infection in 1,016 of them.[12] Multiple similar studies carried out in the US and Europe have found similar results. A fairly average misdiagnosis rate for other medical problems is 11%.

The CDC's posted case rates echo this finding. In 2022, 63,000 cases of Lyme disease were reported to the CDC by state health departments in the US. This number reflects people who tested positive for Lyme disease in whom the pretest probability that they had Lyme disease was considered high because they had been in a Lyme disease area and had the characteristic symptoms. So these are people with a diagnosis of Lyme disease that meets CDC standards. However, estimates of the number of people diagnosed with Lyme disease in 2022 as assessed by other methods, such as through electronic health records, was found to be in the region of 476,000 cases. This means that, according to doctors' records, more than four hundred thousand people were treated for Lyme disease in 2022 who did not have an officially sanctioned, gold standard diagnosis. This suggests that very high numbers of people with negative tests for *B. burgdorferi* or with a low pretest probability of Lyme disease are still being given a diagnosis of Lyme disease. While false negative results do exist in people with true Lyme disease they do not occur at this high rate.[13] The CDC states that the disparity in these two numbers is likely representative of the number of 'patients who are treated based on clinical suspicion but do not actually have Lyme disease'.[14]

Another conundrum to add to the mix is the number of Lyme disease sufferers in Australia, a country where true Lyme disease is believed not to exist. The blacklegged ticks that are known to carry *Borrelia burgdorferi* have never been found in Australia – the warm, dry climate is unsuitable for them – and the bacterium has never been identified in the ticks that do live there. Consequently, the current position of the Australian government Department of Health is that it is not possible to contract Lyme disease in Australia.[15] But still a large number of people in Australia who have never travelled to a Lyme disease prevalent area have been diagnosed with and are being treated for Lyme disease. The charity Lyme Disease Association of Australia estimates that as many as half a million people have been diagnosed with Lyme disease acquired in Australia, despite scientists claiming this is impossible.[16]

There are three types of Lyme disease diagnosis. Not all are equally subject to the under/misdiagnosis debate. Acute Lyme disease (ALD) refers to the sudden illness that happens shortly after the infected tick bite. It *is* largely a straightforward condition to diagnose and treat and most doctors agree on how that diagnosis and treatment should look. It is a short-term illness manifesting with typical symptoms in people known to have been in a Lyme disease endemic area. Another form of the illness called post-treatment Lyme disease syndrome (PTLDS) manifests as ongoing symptoms after treatment for an acute illness. It is more likely to occur if a person has a very delayed diagnosis or if there is evidence of nervous system involvement in the infection. PTLDS is less well understood than ALD. Nobody knows quite what causes the persistent symptoms and there is some controversy about that. It could be an immune reaction, end organ damage or persistent infection. There is disagreement on whether longer-term antibiotics are appropriate but agreement that persistent symptoms can occur after some cases of ALD and that PTLDS is caused by *B. burgdorferi*.

Chronic Lyme disease, CLD, is the main source of the misdiagnosis figures quoted above and the focus of the Lyme wars. The US National Institute of Allergies and Infectious Diseases defines CLD as a term that is 'used to describe symptoms in people who have *no* clinical or diagnostic evidence of a current or past infection with *B. burgdorferi*'.[17] One article in the *New England Journal of Medicine* describes it as 'a broad array of illnesses or symptom complexes for which there is no reproducible or convincing scientific evidence of any relationship to *Borrelia burgdorferi*'.[18] In other words, what these experts are saying is that CLD is a misdiagnosis and not true Lyme disease. People with CLD are considered not to have an active *B. burgdorferi* infection – not just because of negative tests, but also because they do not have the typical clinical presentation and often have not been to a Lyme endemic area. Most have non-specific symptoms which are potentially attributable to multiple different diagnoses. Many, like Sian, have undergone years of medical investigations receiving no answers and it is likely that many of those have a psychosomatic cause for their symptoms.

So how can a person who has never had a documented tick bite, never had the typical rash and has never been to a Lyme endemic area be given a diagnosis of Lyme disease? How can the misdiagnosis rate be so high?

In medicine, there are always diagnostic grey areas. Finding themselves in one of these, a doctor must make a judgement call, choosing to err on the side of under- or overdiagnosis, depending on what they feel is best for the patient. For example, in my work, if someone has symptoms that *could* be due to epilepsy, but the symptoms are minor, such as occasional blank spells, I will often take my time coming to a firm diagnosis, erring on the side of underdiagnosis until I have incontrovertible evidence to the contrary. I know that a diagnosis of epilepsy is hard to reverse and has serious consequences for a person's life, so I don't rush into it. But, if the symptoms are dangerous, like convulsions, and I'm worried for the person's safety, I might jump in with a diagnosis and treatment before I'm really sure – this time leaning in the overdiagnosis direction. Often, if a doctor thinks their patient's symptoms are not too troubling and may be transient, they might avoid labelling them for a while and just wait to see if the problem resolves spontaneously, as so many do.

There is space to under- and overdiagnose in medicine, a factor which doctors must use responsibly, drawing from guidelines and a good knowledge base. These blurred edges of diagnosis allow doctors a great deal of scope to practice differently to others without necessarily doing anything wrong or being accused of malpractice. Guidelines can be adhered to with varying levels of attention. Clinical signs are inherently subjective. What constitutes a typical Lyme disease rash might be interpreted differently by different doctors. Joint pains and fatigue are two of the most common reasons that people seek medical help and a doctor could choose to place great weight on them to support a diagnosis of Lyme disease or dismiss them if they think they are not out of the ordinary.

Laboratory testing is also subject to variation in practice between labs, resulting in varying levels of positive results. In the case of Lyme disease, in which labs look for antibodies rather than for the bacteria themselves, if they are testing for a very wide range of non-specific antibodies they will return a high rate of positive results. That can be done with little fear of recrimination because a diagnosis is never meant to be based on the test result alone – it is clinical. The lab trusts the doctor to decide if a result is a true positive based on what they know about the patient. But many doctors don't fully understand the tests they're ordering. I can say from a wealth of experience that doctors in receipt of lots of positive results will generally err on the side of assuming most are true positives rather than risk being accused of missing a diagnosis. And a patient desperate for a diagnosis is unlikely to be willing to dismiss a positive diagnostic test just because a doctor suspects it isn't reliable.

Of course, all medical professionals and labs are subject to regulations and oversight so there should be limits to the degree of variation between practices. Neither individual doctors nor institutions can act completely outside the jurisdiction of medical regulators. Most doctors are well-meaning and work hard not to under- or overdiagnose any more than can be avoided. However, there are still plentiful subjective areas to plumb for poorly informed doctors, bad doctors and those unscrupulous enough to wish to do so for personal gain.

Those concerned about overdiagnosis of CLD attribute much of the problem to the exploitation of desperate people looking for answers and to a niche group of professionals profiteering through knowing where the grey areas of diagnosis lie. Singular labs in the US and Germany in particular are responsible for a huge proportion of the positive results for *B. burgdorferi* in patients who had tested negative repeatedly in local labs. Of the dozen or so CLD patients from the UK that I interviewed for this book, every single one received a positive test result from the same private lab in the US after having tested negative in the UK. Similarly, a small number of doctors and private institutions are at the centre of most CLD diagnoses. Like Joseph Jemsek, these individuals and companies are subject to scrutiny and occasionally have their knuckles rapped if they are found to have caused harm to patients, but most continue practising. Chronic Lyme disease manifests as a wide range of common symptoms – tiredness, sleep problems, depression, brain fog, muscle pain, headaches – making it an easily available, catch-all explanation for profit-orientated doctors promising answers to people who just want an answer, any answer.

⋆

Overdiagnosis and misdiagnosis exist, but so does underdiagnosis.

Sylvia's husband Bob contracted *B. burgdorferi* infection in 2008 and it changed both their lives forever.

Bob's illness began with a noticeable lump on the back of his knee. It was not a bull's-eye or spreading rash. It was likely a localised skin reaction to a tick bite, although Bob never saw the tick. His doctor gave him antibiotics, but not for Lyme disease. Bob and Sylvia were due to travel to Greece and the antibiotics were to stave off the possibility of cellulitis while the couple were overseas. Two weeks later, Bob did develop the characteristic bull's-eye rash at the site of the presumed bite. That was the beginning of a systemic illness from which Bob has only partly recovered.

Sylvia is a doctor and considered Lyme disease as a potential diagnosis at the start. Bob had been raking leaves in the couple's garden. She recalled him standing knee deep in damp leaves. The lump was at the back of his knee, which is a typical place for ticks to bite. The couple lived on the edge of South Downs National Park where there are deer and deer ticks known to carry *B. burgdorferi*. Bob asked his doctor for a test for Lyme disease but the result was negative. That was assumed to mean that Bob did not have Lyme disease so his doctor treated the rash as ringworm instead. Sylvia didn't argue.

'He had a rash on the back of his leg like a chrysanthemum and I dismissed it because his doctor dismissed it,' Sylvia told me, sounding close to tears.

After that, Bob's health deteriorated. As he got worse, Sylvia thought more about Bob's rash and wanted to revisit the possibility of Lyme disease, but the negative test for *B. burgdorferi* became a sticking point. When Sylvia phoned the office of a respected, mainstream Lyme disease specialist to make an appointment she was turned away. The specialist would not see patients who did not have a positive test in a standardised lab. Bob was cast adrift. The negative test had become proof to all concerned that Bob did not have Lyme disease but Sylvia could not escape her growing clinical suspicion that he did.

'Getting a diagnosis of Lyme disease for Bob required the level of evidence required in a murder trial. Nobody was willing to make the diagnosis on a balance of clinical probabilities,' Sylvia told me.

A clinical diagnosis based on typical symptoms in someone who lives in a Lyme disease area is considered adequate for diagnosis in the CDC and NICE guidelines, so perhaps the controversy around Lyme disease made some doctors wary to make the diagnosis after Bob's negative test? It certainly felt that way to Sylvia. She was forced to do her own research. Sylvia started reading every academic paper she could on the subject. She attended conferences to hear what scientists had to say. At one conference, she met an expert from the Netherlands and arranged for Bob to have a blood test at a university-affiliated teaching hospital there. They tested for a different strain of *B. burgdorferi* and that test was positive. Bob may have tested negative at his local lab because the test was calibrated to a different strain of bacterium than had infected him or the short course of antibiotics he received before going to Greece may have temporarily dampened the immune response.

Up to this point, Bob had only been given oral antibiotics. Intravenous antibiotics are only recommended in certain circumstances for Lyme disease. After oral treatment failed, with the positive tests, Sylvia felt sure that intravenous antibiotics were required, but they were not easy to access. She couldn't get them from her local doctor, who was worried about the risks of prolonged antibiotics. They may also have felt constrained by the negative test in the UK. With Bob still suffering a great deal, the couple felt forced to look for help from doctors whom Sylvia would never have expected to consult in normal circumstances.

'It's like the Lyme wild west out there,' Sylvia said. 'It's impossible to know who to trust. We went to see people I would normally consider to be quacks.'

Bob booked into a UK-based private clinic that is notorious for offering highly unconventional, off-licence treatment that many doctors would consider dangerous.

'I had to go along with their ethos more than I would want to, just to get what we needed,' Sylvia admitted.

Bob had blood tests the couple did not think were either necessary or entirely legitimate. The samples were sent abroad and returned results that Sylvia, as a doctor, considered uninterpretable and therefore pointless. The couple also agreed to a longer term of intravenous antibiotics than Sylvia thought was strictly necessary. But it worked. Bob got better. Six weeks after the intravenous antibiotics, he was well enough to go on a coastal walk, something he could never have done after he became ill. He didn't make a complete recovery but remains considerably better than he was at the peak of his illness. He hasn't needed any further antibiotics.

Sylvia does not agree with the use of excessive or long-term antibiotics for Lyme disease. Nor does she advocate for sending blood samples to non-standardised labs. She works with a Lyme disease charity to help patients and doctors better understand results from standardised labs. She also campaigns for better acknowledgment of medical uncertainties, so that people who are circumstantially likely to have Lyme disease, like Bob, are not dismissed based solely on a negative blood test.

Tick bites are easily missed. Since the typical rash doesn't last long and appears in armpits and on areas of the body that people don't necessarily see or examine on a daily basis it can go unnoticed. The tests used to help make a diagnosis of Lyme disease have never been and are still not perfect. It is also undoubtedly the case that the overdiagnosis of Lyme disease is so notorious among doctors that they doubt every borderline diagnosis they encounter. Unfortunately, the so-called 'Lyme Wars' are harming patients in more than one way. It has created an atmosphere around the disorder that sees doctors distance themselves from patients with all forms of Lyme disease and has probably also contributed to the lack of research into PTLDS. Controversial areas of medicine often find themselves poorly served by research and treatment facilities. There are so many unanswered medical questions that there is no need for any doctor or scientist to choose a controversial topic over others that are less controversial.

Sylvia is also concerned that patients' and researchers' interests do not always align, which leaves patients in the cold. That was something that also worried Polly Murray. In her archive, tucked among family medical records and newspaper clippings, is a 1985 medical paper published in the *Annals of Internal Medicine* called 'An Introduction to Medical Phenomenology: I can't hear you while I'm listening'. In it, the physician Richard Baron describes how pathology comes to define a disease more so than the patient experience. Perhaps once a test was developed to help diagnose Lyme disease, it eclipsed the patient's story.

So underdiagnosis of Lyme disease does exist but, still, the evidence points to overdiagnosis being a much bigger problem. Misdiagnosis flourishes in large part because it has proven so profitable for a small number of unscrupulous practitioners and institutions. But it also grows through serving an important social purpose. Worldwide, there are a great many people suffering without an explanation. Many people do not feel well served by mainstream medicine and profiteers fill that care vacuum. As a disorder that affects multiple different bodily systems, Lyme disease makes for a feasible explanation for a wide range of types of suffering. CLD is a successful diagnosis because it offers everything a diagnosis should – it is applicable to a great number of people, comes with treatment, a community – and, again, hope.

Speaking to a young woman with CLD who had paid a great deal of money to private labs and doctors over several years, I asked for the one message she wanted me to put across on her behalf. We had been talking for some time and I had visibly balked when I learned she had spent tens of thousands of pounds on highly unconventional and potentially dangerous treatments. It was money she could ill afford to spend.

'I want other doctors to learn from my private doctors,' she told me.

I balked again, I admit. It was not the message I expected or cared to pass on. But I have thought about it a great deal since and realised she was absolutely right. This young woman told me that when she went to see her doctor in a private clinic known for controversial, off-licence treatments, the private doctor did numerous tests that nobody else had done. That had inspired immediate confidence. It took me a long time to realise that the 'tests' she was referring to were neither blood tests nor scans. She was referring to the fact that the doctor had spent a great deal of time doing a detailed physical examination. They had listened to her properly, paid attention and laid hands on her. It made me think of Sian who told me that Dr Jemsek encourages her to list all her symptoms every time they speak.

It is tempting to simply criticise doctors whose practice seems to exploit desperate people and medicine's loopholes. But that ignores the inadequacies in mainstream medicine that drive people to these practitioners. There is no place in mainstream medicine for people with multi-system complaints that don't fit neatly into a diagnostic category. The highly specialised nature of medicine sees these patients shunted from doctor to doctor, with each ruling out disease in their category and then considering their obligation to the patient complete. Better long-term care for people without a diagnosis is more likely to make a questionable Lyme specialist's lucrative practice non-viable than the occasional restrictions by regulatory bodies that have, to date, done nothing to meaningfully affect those doctors' practices.

When patients come to see me with seizures and various neurological complaints, I often arrange for them to have a series of extremely sophisticated tests. But I always tell them that the tests are not the most important part of our journey to their diagnosis. That lies much more in my ability to hear and fully understand everything that has happened to bring them to me. Their trust in whichever diagnosis I make will also depend on how well I have done that listening part of my job. It can take several meetings to get there. Diagnosis often happens in stages as doctor and patient become more relaxed in each other's presence and the full story slowly unravels.

Lyme disease diagnosis is just one example of the malleable nature of medical diagnoses and of diagnostic tests. It is a reminder of the need for clinical context. But more importantly to this book, perhaps, it also reveals that when a diagnosis makes people feel better, it isn't necessarily because of the specifics of the diagnosis itself or the treatment, but because of how it makes a person feel heard. And it is just that validating quality of a diagnosis, its ability to make people feel better through being believed, that is the major driving force of the rising rates of many other diagnoses too.

*

Although uncovering the cause of Lyme disease was collaborative between Polly and the Connecticut State Health

Department, she didn't agree with the researchers' early descriptions of the disease, which focused heavily on arthritis as a central feature, meaning the disorder was first called Lyme arthritis. That didn't resonate with Polly's wide-ranging symptoms. Hers was a multi-system disease. To convince researchers of such, long after the scientists had taken over, she continued to document her symptoms and those of her neighbours. And she was proven right. As the researchers learned more, they acknowledged that multiple organs could be affected and Lyme arthritis became Lyme disease.

The Yale library holds an archive of her research, where I spent time poring over boxes filled with her notes, letters, photographs and newspaper clippings. There are pages and pages of lists of symptoms – rash, hip pain, insomnia, sore throat, night sweats, cough, diarrhoea, seizures, stuttering, hiccups, tremor, sudden fear of heights, change in handwriting, mouth clicking open, lump when swallowing, double vision, nail pitting, bad breath, aggressive behaviour, difficulty making decisions and so on. In documenting every unexplained symptom she came across, Polly will inevitably have drawn people under the Lyme disease umbrella who didn't belong there.

Disease descriptions can grow very quickly if they have no immutable defining characteristic or reliable diagnostic test. When a new medical diagnosis comes along, it presents a potential explanation for people with previously undiagnosed problems. Patients want answers and doctors want to provide them so a new inclusive diagnosis associated with lots of common symptoms is always welcome. The array of symptoms associated with a novel condition can expand very quickly because every new person identified as having the diagnosis will bring features from their own personal experience to it.

Lyme disease started as arthritis, welcoming people with so-far-unexplained arthritis into the diagnostic classification.

With a bull's-eye rash added to the disease description, people with this rash, with or without arthritis, could also be given the diagnosis. Over time, people with a more nondescript joint problem and less specific type of rash entered the group. As new people were diagnosed, the symptoms introduced to the classification by new sufferers became less specific – such as fatigue, difficulty concentrating and poor memory. Until inevitably, some later members of the group did not have either of the characteristic first symptoms that defined the condition at the start – arthritis or a bull's-eye rash. It became possible for unexplained fatigue or memory problems to be enough to qualify for the diagnosis. This is how a diagnosis can evolve over time if its symptoms are somewhat vague and it doesn't have a well-defined clinical presentation or reliable test to limit it. Years after a new disease has been coined, those diagnosed may look nothing at all like those with the classical features of the disease when first described. This is something we will encounter again and again in later chapters, particularly with mental health and behavioural conditions like autism and ADHD.

But new people don't just change the features of the diagnosis by bringing new symptoms of their own to it, they are also changed by the act of being labelled. In just the same way that Valentina was certain she had symptoms of HD because she knew she was at risk, a person given a diagnosis like CLD may start to develop an array of symptoms known to be associated with CLD that they didn't have before they were formally diagnosed.

So, a person might start with one unexplained symptom like a flitting rash. A new diagnosis called CLD comes into being. It becomes part of the doctor's diagnostic armoury for when they encounter people with odd, hard-to-diagnose rashes. The person without a diagnosis is told they might have CLD. They are advised of other common CLD symptoms and, quite naturally, look for them. The body processes that normally filter out bodily white noise are disrupted through paying excessive attention and the person begins to notice new symptoms that seem to fit with Lyme disease. But actually, those new symptoms are psychosomatic and have arisen out of expectations and predictive coding. In that way, a person who didn't start out with very typical symptoms of Lyme disease might develop them over time through the nocebo effect of labelling.

The widening disease description of Lyme and the increasing number of diagnoses from the late 1980s onwards was facilitated by the fact that the bacteria are hard to discover in the human body and the disease has such wide-ranging symptoms. However, it was also undoubtedly accelerated by the media's interest in the story of a mother's fight against the system. For a period, Lyme disease was something of a disease-du-jour. One *New York Times* article, published in 1989, said Lyme disease was subject to 'more publicity than Princess Di or Roseanne Barr'. At any point in time, there are always legions of frustrated, even desperate people with unexplained physical symptoms. People related to Polly's struggle to be diagnosed, until even those who did not live in Lyme endemic areas believed they had what she had.

*

Usually, doctors and scientists hold all the power in developing new diagnostic concepts. However, like Lyme disease, a diagnosis that began with the observations of patients is long Covid. The term was coined by Elisa Perego when, on 20 May 2020, she used it as a hashtag on Twitter to draw together multiple threads of concern about the long-term consequences of Covid-19. Perego lived in Italy, where the pandemic had taken off with particular ferocity after the first case was reported in mid-February 2020. She was infected soon after. Although her acute infection was not severe, she struggled to recover. She was one of many who turned to social media to report the possibility of persistent symptoms after an apparently mild infection. The traditional media's focus at the time was on those who were dying, but social media was bringing to light a whole other population who were not in hospital but who were also suffering. Perego invited others who had not recovered after a mild infection to share their experiences on social media using #longcovid. Her campaign took off very quickly. People came forward in their droves to tell their stories. By June 2020, many were identifying as '#longhaulers', predicting chronic disability for millions.

The reports of persistent symptoms after a mild infection alongside the term long Covid spread quickly from social media to news programmes and newspapers and then to medical literature. In June 2020, in the *Atlantic*, science writer Ed Yong described young people attacked by rolling waves of symptoms from which they could not recover.[19] In July, the *British Medical Journal* described an influx of long Covid to GP practices.[20] Soon after, politicians began to present the public with this fresh threat as another reason as to why they must stay isolated to keep themselves safe.

Much of the early conversation about long Covid was driven by patient-activists and patient-scientists. New Yorker Fiona Lowenstein contracted Covid-19 in March 2020. Although her acute illness was short-lived and fairly typical, it later morphed into symptoms not previously associated with Covid, such as hives and sinus pain. Lowenstein went on to set up the Body Politic Covid-19 support group.[21] [22] Their mission was to facilitate patient-led and patient-involved research into Covid-19. They wanted a paradigm shift in which patients served as leaders in their own care. They referred to their movement as 'health democratised'.

In the UK too, patients were at the forefront of the drive to have long Covid taken seriously. In an open letter published in the *British Medical Journal* in September 2020, Dr Elisa Perego, who is a researcher in archaeology rather than a medical scientist, described long Covid as a 'cyclical, multiphasic and multi-system condition'.[23] Her co-author, Nisreen Alwan, a public health researcher and another long Covid patient, became a staple member of many public panels on the subject. Together, Perego and Alwan reinforced the message that long Covid was patient-created and that, going forward, patient experts should be central to research efforts.[24]

From the outset, long Covid – as a diagnostic concept, at least – was problematic. Alwan, as one of the pioneer doctors involved in supporting people with long Covid, defined the condition as 'not recovering several weeks or months after you had symptoms suggestive of Covid whether you were tested or not'. There was no disease definition. There were no specific symptoms and no diagnostic test results to set limits to the diagnosis. It was usually a self-diagnosis and no proof of infection was required. A negative Covid-19 test did not count against it. In fact, a substantial number of people with long Covid then and now either tested negative for the virus or had not been tested.[25] When polled early in the pandemic, 70% of the members of one UK-based long Covid support group had tested negative for the virus. In an early survey run by Lowenstein's Body Politic, as few as 15.9% of those with long Covid had tested positive for the virus.[26]

That there was no disease definition saw the list of symptoms attributed to long Covid grow to more than 200 within days. Many of the symptoms were so non-specific that it would be hard to find a person who did not have at least one – such as dizziness, tiredness, depression, anxiety, muscle pain, headaches, nausea, joint pain, difficulty sleeping. One support group included 'loneliness', 'feeling scared' and 'skin ageing' on their symptom list. At one public seminar I attended, Alwan asked people to consider consulting their doctor even if they had *no* symptoms, for fear there could be people who had long Covid without realising it. If a person didn't need symptoms or a positive Covid test, that meant anyone or everyone could have it. With no diagnostic criteria to define it, long Covid could now feasibly be the cause of all sorts of physical and psychological suffering during the pandemic.

A great deal about long Covid as a single discrete condition caused by a viral infection was counterintuitive. With most infectious illnesses, or any disease really, the sicker a person is at the peak of their illness the less likely they are to recover and the more likely they are to have lingering medical problems. Yet long Covid's relationship to an acute Covid-19 infection was the opposite to that. Long Covid was more common after a mild infection than in those hospitalised with severe disease.[27] Sicker people who survived the acute infection were making more complete recoveries than those with a mild infection. Some studies showed that non-hospitalised long Covid patients had a wider range of more severe symptoms than hospitalised patients.[28 29]

The demographic of those with long Covid was also different to those hospitalised.[30] Patients admitted to intensive care units and who died from Covid were more often the elderly, male and those with comorbidities like diabetes. But long Covid patients were more likely to be younger and female, and diabetes was not a risk factor. Long Covid patients in the media often described good health and exceptional fitness pre-pandemic – the opposite of hospitalised patients, who were often older and frail. And, while acute Covid had some quite well-circumscribed symptoms, the manifestations of long Covid were highly variable and prone to change. One sufferer described them as 'protean – imagine a mischief of mice moving through the walls of your house and laying waste to different bits of circuitry and infrastructure as they go.'[31]

Years later, under the name 'post Covid-19', long Covid is defined by the WHO (World Health Organisation) as the continuation or development of new symptoms three months after the initial SARS-CoV-2 infection, with these symptoms lasting for at least two months with no other explanation. The WHO estimates that 17 million people might have been affected in Europe in the first two years of the pandemic alone.[32] Approximately 1.9 million in the UK had self-assessed with the disorder by March 2023.[33] In March 2024 nearly 18 million American adults reported ongoing long Covid symptoms.[34]

A variety of medical problems have undoubtedly been drawn under the single label of long Covid. They fall into four main groups.

Firstly, persistent symptoms would not be surprising for anybody sick enough to be hospitalised with Covid-19. I spent some time volunteering in the intensive care unit during the pandemic and I saw the toll the virus took. In the hospitalised, especially those who required support for their breathing, persistent symptoms could arise as a result of organ damage caused by the virus, side effects of treatment and the complications of a long hospital stay.

Secondly, a percentage of those not hospitalised with Covid must certainly have developed persistent symptoms as part of a post-viral fatigue syndrome. This is a well-documented but poorly understood consequence after many viral infections. Most post-viral fatigue syndromes recover spontaneously, although they can be very severe and disabling while they last.

Thirdly, the lack of face-to-face appointments and overwhelmed healthcare facilities will certainly have seen some people misdiagnosed as either having Covid-19 or long Covid at the peak of the outbreak. In the middle of a pandemic, it is very easy to blame everything on the virus and to miss ordinary medical problems. Case reports are already emerging of people who were told they had Covid only to later discover that they had another medical problem, including cancer.

However, these three explanations are unlikely to account for the largest group diagnosed with long Covid – those with a mild infection, often with a negative test for Covid or self-diagnosed, who have a wide range of persistent symptoms beyond those typically associated with the acute infection. The symptoms that this fourth group describe are very difficult to explain through pathological mechanisms related directly to a viral infection and are best explained by a psychosomatic illness.[35] [36] [37] [38]

The evidence that a significant proportion of long Covid has a psychosomatic cause has built slowly. Multiple studies have shown that anxiety, depression and perceived stress are consistent risk factors for long Covid.[39] [40] [41] [42] [43] One Norwegian study found the presence of loneliness or negative life events in the year prior to contracting Covid was a better predicter of who would develop long Covid than a positive test for the virus.[44] A UK-based study of more than 30,000 children and young people also identified loneliness as closely associated with the development of long Covid.[45] A German study that followed healthcare workers through the pandemic showed that psychosocial burden and expectations of symptom severity were risk factors for long Covid.[46] A French study demonstrated that self-reported Covid-19 infection was more likely to lead to long Covid than laboratory confirmed infection. Those who expected to get sick, did.[47] [48]

Long Covid behaves just as psychosomatic illnesses do, with a flitting myriad of symptoms that defy anatomical explanation. Because there is no defining pathology in any single organ, it involves different bodily systems in different combinations in different people. Non-hospitalised long Covid patients often had a wider range of more severe symptoms than those recovering from hospitalisation. It consistently contradicts biology. For example, sufferers who are the most short of breath also have the most normal lung function tests and medical investigations do not explain and are often at odds with the symptoms.[49] [50]

Long Covid as a psychosomatic disorder could have arisen through a variety of mechanisms. As we've seen, the nocebo effect is a potent generator of physical symptoms through the power of belief. Excess attention paid to the body during the pandemic changed how people experienced and used their body. Searching the body for evidence of infection may then have brought existing symptoms that might previously have been dismissed to the fore. Adrenaline, a change in diet and/or alcohol intake, and different activity levels negatively impacted general health and increased bodily white noise. Predictive coding – the means by which our brains process bodily sensations through the framework of our expectations – used our vivid internal models of the havoc Covid-19 could wreak and flooded our bodies with false signals. Some people's expectation of long-term illness may have come to outweigh the reality of their healthy bodies as horror stories bore down on them from all sides. Long Covid spread through fear as alarming news and misinformation flooded mainstream and social media. Rather than being a single illness, long Covid was the array of consequences one can expect when people are caught in a global pandemic and have nowhere to turn.[51]

Attempts to explain all long Covid as a consequence of viral pathology have so far fallen short. The presence of small clots in the blood, known as 'microclots', are often mooted as a cause and a potential route to treatment. That theory has been investigated by a Cochrane review, a systematic assessment of all available research results on a particular topic, and no evidence was found to support this. Studies found that microclots are equally found in other diseases and in healthy people.[52]

Pathology that potentially explains persistent symptoms *has* been found in hospitalised patients who had severe infections. The virus really took a physical toll on those hit hardest and that has been reflected in the inflammation and ongoing infection seen in some of those who were hospitalised and in some of those who, sadly, died. But these findings are often assumed to apply to all the other long Covid groups equally and are used to argue against a psychosomatic cause – when actually there is no evidence to say that they should apply to all groups. Normal tests with no evidence for infection or inflammation are the consistent finding in the group in whom psychosomatic illness makes best clinical sense. The most defining characteristic of those with long Covid after a mild or self-diagnosed infection is the lack of association between symptoms and tests and the lack of evidence for infection and inflammation.[53] [54] One study showed a greater correlation between psychiatric symptoms and disordered breathing than with lung pathology.[55] It is also worth noting that raised inflammatory markers are seen in association with stress and mental health conditions; so, if they did exist, they could also be explained in ways other than as a consequence of infection.

In fact, the rush to explain long Covid saw patients who probably had multiple different problems amassed as a group, which has made the likelihood of finding a commonality between them very unlikely. Many early studies that showed a wide range of persistent symptoms after an acute Covid-19 infection were of low quality and did not even include a control group. Later studies showed just as many symptoms in control subjects as were seen in people with long Covid. This suggests that the symptoms of long Covid were not exclusive to those infected and were more likely to be caused by some aspect of the social restrictions in the pandemic rather than by the virus itself.[56]

It's worth noting that the manifestations of long Covid and chronic Lyme disease are very similar. A person might be given either diagnosis depending on when and where they got sick. Both commonly come with a lot of non-specific symptoms such as fatigue, brain fog, depression, aches and pains, sleep problems, dizziness. Both disorders are so ill-defined that the number of symptoms associated with them is far more than those associated with either proven acute Covid-19 or true Lyme disease.

It is very hard to talk openly about psychological mechanisms as being responsible for either of these conditions. I know that conversation will upset people, partly because it will be misunderstood. Psychosomatic disorders are often confused with malingering when the two are completely unrelated. Malingering is deliberately pretending to be ill. But psychosomatic disorders are disabling, unconsciously generated medical disorders. Someone diagnosed with a psychosomatic disorder may feel they are being dismissed by the medical establishment, as, in the popular imagination, psychosomatic disorders are seen as 'lesser' than other diseases, which means the diagnosis is taken to diminish the person's level of suffering. Many people still assume that they are imagined or purposeful.

None of that is true. Psychosomatic disorders produce *real* physical symptoms. Like palpitations that occur with a sudden fright, psychosomatic symptoms are genuinely experienced but are not due to a disease. A psychosomatic mechanism does not mean a person isn't disabled, only that the problem has arisen through a complex mind–body interaction rather than being due to tissue pathology caused by a virus. I have spent much of my career making people aware of the seriousness of psychosomatic conditions. I can assure you the pain or fatigue of a psychosomatic condition can be as debilitating as that of cancer and this explanation should not be seen to diminish the suffering of those with this form of long Covid.

It is precisely because it is such a difficult conversation that a psychosomatic explanation for people with long Covid has not featured nearly enough in public discourse. Even doctors and scientists who strongly suspect long Covid and CLD are psychosomatic don't necessarily raise it. That fear has stymied research progress into both. When one scientist published the results of her research into long Covid, which showed that people who were under stress were more likely to develop long Covid, she attached a disclaimer to the publication saying that this should not be 'misinterpreted as supporting a hypothesis that post-COVID-19 conditions are psychosomatic'. She later expressed regret for that statement, saying that she meant only that people with long Covid were not pretending or duping their doctors.[57] [58] Her research *did* in fact lend support to long Covid being psychosomatic, based on the technical definition of the term, she said later. It seems she was afraid to say that in the first instance knowing the misunderstandings it could provoke.

Patients often reject this explanation for their suffering as psychosomatic symptoms are all too commonly presented as less disabling than the symptoms of other medical problems. Psychosomatic, psychological and social suffering are still highly stigmatised and neglected. Disorders like long Covid and CLD present more appealing alternatives to those people with multiple disabling symptoms who are looking for a diagnosis and treatment, as they allow people to access help while navigating a way around stigma. They also offer a support network, peace of mind through an explanation and validation that the suffering is real.

Society has a general lack of caring institutions, except for medical facilities. This means that physical illness is always prioritised and so it is more straightforward when distress is expressed as a medical problem. No institution took responsibility for Sian as long as she seemed undiagnosable. There were no systems of support for people stuck at home during the pandemic who did not catch the virus. It wasn't easy for a person with unexplained symptoms to visit their doctor, which meant people missed out on vital reassurance. As Elisa Perego's campaign pointed out, quite rightly, it was not only those who were severely ill who were suffering. A medical diagnosis brings a person under the jurisdiction of one of the few institutions available to offer support in a crisis. Without long Covid, many of those feeling the social and psychological effect of the pandemic had no voice.

Disorders like CLD and long Covid grow in prevalence because they bring help to people who need help. That is facilitated by the fact that the subjectivity of diagnosis makes it so available.

*

I am a highly specialised doctor who works in a state-of-the-art environment. I understand the value of tests and technology when they're used well. I don't want to minimise the importance of the technical diagnostic tools that I have to hand today that did not exist, or were not as reliable, when I qualified as a doctor. Back then, before magnetic resonance imaging (MRI) was widely available, it took months, or even years, of unpleasant, invasive tests to make a diagnosis of a disease like multiple sclerosis. I remember so many people in comas, with seizures, paralysis and other mystery illnesses that I couldn't explain back then but could easily explain today through scientific advances. ELISA was developed in the 1970s and Western blot in the 1980s. They are an essential part of a doctor's diagnostic toolbox and a key research tool. They give confidence and speed to a diagnosis.

That the tests used to help diagnose Lyme disease can be tricky to interpret should not be seen to denigrate the extreme usefulness of tests but should rather help people to understand that tests alone cannot themselves make a diagnosis. Used alone, they can be uninterpretable and even misleading. Diagnosis is an art and a science, but, surprising as it may sound to some, the art still takes precedence. All tests have a false positive and a false negative rate, so a test that is done as a kind of sweeping diagnostic survey in a person with vague symptoms has a high chance of producing confusing results. If a doctor doesn't know all the confounding variables that affect the test they have requested, they can easily give too much weight to an abnormal, or a normal, result. Frankly, it's impossible for any doctor to know all the confounding variables for every test, so those requesting tests they don't encounter every week can easily find themselves out of their depth. And doctors working with tests they *do* order every week can find themselves out of their depth too! As a neurologist, I send patients for MRI brain scans countless times a month, but I wouldn't claim to know the meaning of every single tentatively 'abnormal' finding. I am regularly handed a scan that shows something that may or may not be relevant to my patient. When that happens, only a discussion with colleagues and radiologists in which the patient's story is given central position will resolve the dilemma.

Nor do I want to imply that normal tests, as are seen in many people with CLD and long Covid, are the whole reason that a doctor says a disorder could be psychosomatic. They aren't. There are lots of diseases that are hard to detect with tests. A large number of the patients I see with epilepsy have normal test results, but the diagnosis of epilepsy can still be made with some confidence based on the description of the seizures.

The psychosomatic diagnosis is made in just the same way – based on how the symptoms present, the way they evolve, the way they move around the body, defying anatomy, and also on the contradiction between physical findings and the degree of symptoms.

Diagnosis is made through inferences, using tests in concert with all the clinical information. It is a clinical art that is highly qualitative and cannot be distilled to a simple analysis based on a list of common symptoms. Doctors are prepared to dismiss positive and negative tests equally, if the clinical diagnosis is strong. An excellent diagnostician is one with good clinical acumen and, more often than not, they are also the one with the most experience. Doctors continue to learn throughout their career through listening to their patient's subjective account of their symptoms. I find these conversations fascinating. They speak to how unlikely it is that feeding symptoms into a computer will ever produce a reliable diagnosis. If in talking about a patient's headache I use the word 'pain', they might correct me and say, no, it's not a pain, it's a 'soreness'. Unpicking why one word is a better description than another similar word is often the most important part of a diagnostic discussion. Medical investigations contribute to diagnosis, but they are not the objective thing they are perceived to be. They can be misused, misinterpreted and manipulated. The quality of the doctor doing the test is as important, if not more important, than the quality of the test.

Developing new diagnoses is also a science and an art, but in this instance, the science must take precedence. In seeking help for people with long Covid, Fiona Lowenstein's Body Politic movement called for the democratisation of health. But scientific answers aren't at the convenience of the majority opinion. Understanding patient experience is fundamental to setting research priorities, but scientific process must still be systematic, methodical, rigorous and open to any answer. Both Lyme disease and long Covid gained such traction in the public arena that they grew outside of the control of the researchers. The discussion around long Covid moved from social media to mainstream media to medical media without adequate scientific interrogation. Society's demand for immediate answers didn't allow enough time for scientists to stand back and ask the most basic question of long Covid: how do we reliably define this new disorder so we can study it methodically? Social pressure during the pandemic led scientists and doctors to skip over the essential work of dividing up different types of long Covid for individual study to make research and conversation around the disorder more meaningful. The public rejection of a psychosomatic formulation to explain long Covid shut down that important line of enquiry to the detriment of those affected.

A good doctor is one who is experienced enough to hear the nuance in the story, who understands the primacy of the clinical context, who doesn't order tests every time, who doesn't give a label to every ailment, who knows when watchful waiting is the best strategy but also knows when to act. A good scientist is one who is logical and careful, who is objective, who is creative and playful with ideas, but who is also sceptical and unafraid to question strongly held assumptions.

3
Autism

Poppy is twenty-four years old. She presents as confident, although I know that isn't easy for her. She has packed more than the average amount of difficulties into her young life and has only lately started to feel better. Her recent diagnosis of autism rescued her from a crisis point in her life.

'Lots of people seek to be diagnosed with autism but it wasn't like that for me,' Poppy told me. 'I'd been struggling to the point that I thought I couldn't deal with . . . *this* anymore,' she gesticulated around her. 'The only way my brain could handle things was to end it completely.'

She was talking about her life. She had seriously considered suicide twice before and, aged twenty, she was there again.

Poppy's mental health problems began when she was twelve, first with depression leading to self-harm. In her early teens, she developed an eating disorder which she has only partially got under control. There were triggers for some of this. At school, she loved the subjects she was studying but struggled to make friends and was bullied relentlessly.

'It started with classic name calling. *She's a weirdo. She's a freak.* That progressed to me being pushed off my scooter, with rocks and food thrown at me. I was targeted in PE. They threw netballs at me. I once had to go home with shattered glasses and a black eye.'

Added to that, life outside school hours was challenging, especially when her dream part-time job turned sour.

'I worked in an aquarium,' Poppy smiled. 'Amphibians are my favourite animals. I have two turtles. I love all animals. I used to have a snake, lizards, rats, mice, hamsters. I loved it there but I had to leave.'

Poppy was very happy at the aquarium to begin with but after two supportive colleagues resigned, she found herself isolated in a work environment that she found very stressful. Her boss, a much older man with medical problems of his own, was very difficult for her to be alone with. To compound her stresses, Poppy found herself in a relationship that was so abusive it led to a police case. Then, against the background of this major trauma, a new job pushed her over the edge. The proverbial straw that broke the camel's back.

After the aquarium, Poppy took the position of showroom host in a high-end car dealership. This certainly wasn't her dream job. The role required her to meet new clients at the door, to bring them coffee and charm them into the showroom. It was completely unsuited to her personality and her talents. She had weathered a lot by then, but with the pressure of the work and unaccommodating bosses she had a breakdown.

'I don't actually remember my meltdown,' Poppy told me. 'My brain blocked it out. I need other people to remind me what happened.'

Poppy's mood plummeted. She couldn't function. That was when she thought about ending her life. I asked what stopped her.

'I was lucky. My girlfriend called me and that snapped me out of it.' Poppy's then girlfriend phoned just by chance, rescuing her. Poppy went on, 'I was reminded of a talk in school about the crisis team and I managed to ring them.'

Poppy has had numerous encounters with mental health services. She was seen by the child and adolescent mental health team (CAMHS) after she first tried to kill herself, aged thirteen. Later, she saw a trauma counsellor and then a psychiatrist. They referred her for EMDR, 'eye movement desensitisation and reprocessing therapy', a technique said to help people process traumatic memories. She found that helpful but after the prescribed course finished, she found herself cast adrift.

'Do you think your mood problems were caused by the bullying or the difficult relationship or a bit of everything?' I asked, still trying to understand the timeline of all that had happened.

'It all happened because nobody noticed I had autism, so I didn't get the help I needed.'

People with autism have higher rates of anorexia and self-harm and are more likely to suffer abuse in relationships than others. For many, the diagnosis of autism only comes after they develop mental health problems. Many professionals in the field believe that those mental health and social problems could be prevented if adjustments for autism were made at an earlier stage. Poppy certainly felt that way.

'Nobody at school picked up on your difficulties?'

'They said it was all just puberty. They acted like what I was going through was just a rite of passage. If just one teacher had noticed a single autistic trait I could have had a much more meaningful school life. Instead, my parents were in the school every other week because of bullying,' she shrugged.

Poppy's school had a special educational needs coordinator and there were autistic pupils in receipt of extra support, but Poppy wasn't recognised as needing the same.

'So how did the suggestion of autism ultimately arise?' I asked.

'My mum actually raised it a long time ago. When I interviewed for the job in the aquarium she sat in on the interview because of my age. Afterwards, she said I hadn't made eye contact once. But back then, I thought autism was something that happened to boys, so I didn't think I had it. After my meltdown, I saw the psychiatrist again. My mum asked him if I might be autistic.'

Poppy's diagnosis was made over the course of a two-hour appointment with a psychiatrist. I asked her how detailed that assessment had been.

'We spoke about my whole life from my point of view and a parent's point of view. I filled out forms and my mum filled out forms. Then I went back two weeks later. The first psychiatrist had left so I saw a different one. They told me I had autism.'

'How did that feel?'

'Actually, I didn't even hear him saying it. I was staring at a geometric picture on the wall. The pattern didn't have a beginning or an end. The psychiatrist told me that he'd hung it there deliberately because if anybody stared at it for long enough he knew they were probably autistic.'

For a week after the assessment, Poppy didn't think about the diagnosis. She didn't research it or read about it.

'But at the end of the week, I decided I wanted to know more. I wanted to know if there were other people like me out there, so I went on TikTok and I found that there were lots.'

Reading other people's account of autism on social media, Poppy quickly came around to the view that her diagnosis was correct. It reframed and explained every difficulty she had in her life. The realisation came as a great relief.

'People tell me all the time that I don't look autistic. But what is being autistic supposed to look like? I've been masking all my life.'

Poppy had read my thoughts. In our conversation, I found her to be very engaging without any immediate evidence of communication problems. Although, of course, she was in her comfort zone talking about a subject that possesses her – autism. I am no stranger to appearing confident in some situations and socially inept in others, so I could understand that her apparent comfort with me did not reflect all her sides.

Poppy explained her ability to seem socially confident through 'masking', which refers to a strategy used by autistic people to cover up their typical 'autistic' behaviours so that they can better fit in socially. It's a way of mimicking non-autistic people in an attempt to appear to be 'normal'. I asked Poppy what I would see if she wasn't masking.

'I cannot tell what a neurotypical person is thinking. I don't understand sarcasm.'

Poppy thinks in black and white. She has lots of interests in which she can become deeply involved, including books, art, music and plants. She cannot tolerate loud noise or certain textures or shapes. She struggles with friends, managing better with one good friend rather than groups of friends. Poppy felt that the autism diagnosis was absolutely necessary for her to manage these aspects of herself.

'Without autism, I wouldn't know I'm not lazy. I know now that I can't physically make myself do things because I have the wrong dopamine. Without autism, I would never have learned that I have pathological demand avoidance. I would never have learned that there's a reason that I have all the things that I have been so insecure about my entire life.'

One theory is that autism is caused by low levels of dopamine in the brain. This is a neurotransmitter associated with reward and pleasure. Pathological demand avoidance (PDA) isn't included in any official description of autism but is increasingly spoken of as part of the diagnosis. It refers to an autistic person's difficulty complying with requests. The UK's National Autistic Society gives examples of PDA that include a child resolutely refusing to brush their teeth or to put on their coat, or a person failing to respond to their own body's demands, such as the need to eat. Clearly, these are behaviours many children exhibit, but the implication is that for autistic people they are insurmountable.

Since her diagnosis, Poppy hasn't had any specific medical intervention for autism but is a great deal happier. She feels better able to manage her difficulties because she recognises where they come from. She unmasks when she can, allowing herself to be herself. Currently, she's a freelance worker, giving talks about autism and being paid for social media content. She has still to fulfil her dream of becoming a marine biologist. She had a place on a foundation course for marine biology but that didn't go as planned and she was forced to leave. When she applied, the college promised they had support systems in place for people with autism, but she hadn't felt supported. The commute to the college was too long so she left before completing first year. She would like to try again, but, until then, college has been replaced by lots of new friends in the autism community. Albeit mostly online.

I asked what support she'd been offered by the NHS.

'I didn't even get a leaflet, but I'm glad I didn't. There is a lot of harmful information that people in the NHS and special education system still hold about autism.'

'You're not happy with the NHS message on autism?'

'Too much misinformation.'

'For example?'

'The function labels for starters. If you call someone high functioning it's like you're saying they're not disabled enough. And don't call me a person *with* autism. That sounds like I'm carrying autism around in a bag. I'm autistic,' Poppy advised me. 'That's how my brain works. It's part of who I am. It's not something that can be put down like a bag.'

Autistic people are often referred to as 'low' functioning or 'high' functioning, with the latter referring to those with normal or above average intellect and the former referring to those with an intellectual disability. Some sections of the autism community find the distinction insulting because of how it appears to diminish the level of difficulties faced by those said to be high functioning. For the same reason, the concept of a 'spectrum' and the use of the term 'mild autism' are strongly rejected by some.

The labels used to refer to people with an autism diagnosis are also contentious. Some prefer the term 'people with autism', thus separating the disorder from the identity of the person. That is the convention with most medical problems. People with epilepsy and diabetes don't usually want to be referred to as epileptics and diabetics because they don't want to be defined by their disease. But in the case of autism, a growing number of those like Poppy, especially those who would traditionally fall in the category of 'high functioning' and 'mild', actually prefer the term 'autistic person', precisely because they consider being autistic as integral to who they are.

Since her diagnosis Poppy has made lots of videos about her experience of autism which she posts on social media. She has attracted a large following of people with similar problems, the majority of whom are in her age group and have what would usually be referred to as mild autism. The majority of those also disagree with the use of the term 'severe autism' to refer to people who are too disabled to live independently.

As Poppy explained it to me, 'Somebody classed as low functioning or severely autistic may be capable of much more independence than anyone thinks. Their freedom to make choices is stripped away from them just because of the label. They could lose a lot of things they might have enjoyed in life because the label makes them sound as if they're not capable.'

'Couldn't that argument also be used to mean that you are underestimating *your* abilities and limiting *your* future by taking on the label "autistic" . . . ?' I ventured.

'I know some people say calling a person autistic will hold them back but I don't agree with that,' Poppy said emphatically, not realising she had contradicted herself.

In support of Poppy's view of her own diagnosis, I have to agree that so far at least, autism has opened up her world and helped her keep an even keel.

I hesitated before asking something that had been bothering me since the beginning of our conversation. 'I've been wondering, Poppy, if your struggle to manage might not be better thought of as a consequence of some of the things you've lived through, rather than being due to a neurodevelopmental brain disorder?'

'I agree with that, actually. If I hadn't gone through what I'd gone through, I'd be much happier. I knew I wasn't normal and other people picked up on that.'

Poppy attributes some of her ongoing difficulties to her traumatic experiences, but at the same time thinks that she would not have been subjected to those traumatic experiences if she was not innately neurodevelopmentally different.

★

Most people in the developed world will have noticed the rising number of people with a diagnosis of autism. One in thirty-six American children now has autism, up from 1 in 150 twenty years ago.[1] But how has that increase come about? Is it, as some believe, evidence that we are getting better at making the diagnosis? Or, as others suspect, an example of overdiagnosis? If you ask this question of specialists in the field, even at the highest level, you might be given contradictory answers. 'Pure fantasy', is how highly experienced Canadian autism specialist Laurent Mottron referred to the assertion that one in five children in Northern Ireland is autistic.[2] Mottron's view of the worldwide rise in autism prevalence is that, 'the condition is being diagnosed at a rate that's simply not consistent with reality.'[3] But referring to the rising rate of the diagnosis in the wider UK, where one in thirty-six children is affected, equally experienced autism researcher Simon Baron-Cohen expressed no concern about overdiagnosis, saying, 'I think the rates we see today are getting closer to the true rate.'[4] If Mottron is correct then it may be that people are being labelled inappropriately with a neurodevelopmental brain disorder, which is important to know, given the impact of a diagnosis on a person's sense of self. If Baron-Cohen's view is closer to the truth, then are we starting to solve the problem of discrimination against children who learn and socialise differently?

There is no blood test or scan to diagnose autism. The diagnosis depends entirely on a societal agreement on what normal behaviour should look like, the observation of abnormal behaviour in an individual and that person's description of their internal experience. Diagnosis in general is fraught with error even when objective tests are involved, but now we enter the quagmire of trying to define and diagnose a condition for which there are no tests and absolutely no objective clinical signs.

Both defining and diagnosing mental health and neurodevelopmental disorders has always been a challenge. The *Diagnostic and Statistical Manual of Mental Disorders* (DSM), first published in 1952, was introduced in a direct attempt to address this. The DSM describes typical clinical presentations and diagnostic criteria for all mental health disorders. The DSM-5, published in 2013, is the seventh and latest edition of this encyclopaedia of psychiatric and psychological conditions.[5]

The diagnostic criteria for autism set out by the DSM includes two areas of difficulties. Firstly, problems with social communication and interaction. Secondly, restricted or repetitive interests and patterns of behaviour. To have autism, a child must have problems in both these areas and they must have been evident since the early developmental period, meaning before the age of five. There must also be evidence of impairment, which is easy to spot at the severe end of the spectrum, but much less well defined at the mild end of the spectrum.

Under the heading of social and communication difficulties, the DSM describes severe-to-mild impairment as ranging from a child who is non-verbal to one with normal language development but who struggles to maintain back-and-forth conversation. Children with autism also tend to use few hand gestures. They often cannot appreciate what others may be feeling. Poor eye contact is common, as is a lack of interest in others and a preference for being alone. These difficulties shouldn't only be evident in situations that might challenge lots of people or in a single context such as school, but should also be apparent in familiar places like home.

Restrictive and repetitive patterns of behaviour can manifest in various ways. A child may have a strong attachment to an object or an unusual interest in or aversion to certain textures. They may have very inflexible personalities. Many exhibit self-stimulatory behaviour (sometimes referred to as 'stims'), such as rocking or bouncing repeatedly. Obsessional interests can take up huge periods of time to the exclusion of all other parts of life. Interests may be highly specific – so not just an interest in cars but in the engine sizes of a particular make of car; not just an interest in the London Underground but in a single line of the Underground. The need for sameness can lead to marked distress if a routine is disrupted.

Although autism is currently subject to an under/overdiagnosis debate, the fact that it was historically underdiagnosed is not really in contention. Until recently, children with autism without a learning disability were often dismissed as stupid or odd because they could not conform to a conventional way of learning or to a traditional school environment. With raised awareness and proactive programmes to seek out autism during school years, most would agree that there is considerably less underdiagnosis now than in the past. The question raised above is: has the correction gone too far or not quite far enough? Those who say we are now overdiagnosing cite the dramatic rise in case rates and the dilution of what it means to be autistic. Between 1998 and 2018, autism diagnoses rose by 787% in the UK. Others applaud the improved diagnosis rates but say that underdiagnosis is ongoing, especially in females and adults.

What is clear is that what autism looks like now is very different to what it looked like when first described. In 1943, child psychiatrist Leo Kanner published a detailed description of eleven children, eight boys and three girls, under his care who he believed had a unique medical disorder manifesting as an 'extreme autistic aloneness'. The characteristic feature was a complete inability to relate in any ordinary way to others. People mattered as much to these children as bookshelves and filing cabinets. They related no better to their own family members and other children than they did to strangers. The children were rigid and obsessive and made monotonous noises and repetitive movements.

The evolution of Kanner's autism to the present-day autistic spectrum disorder began in earnest in the 1960s with the work of Lorna Wing, a psychiatrist and mother of a child with autism. On examining a large cohort of children attending child psychiatric services in Camberwell, London, Wing and colleagues decided that autism was much more common than had previously been suspected and affected people with a variety of intellectual abilities, not just those with severe learning problems. She described a typical triad of impairments that she believed characterised the condition: difficulty with social interaction, communication and imagination. Wing didn't change the fundamental diagnostic features of autism, but she changed the degree to which the typical symptoms had to be present to fulfil the diagnosis, which brought milder cases into the autism fold. She also drew on the work of Hans Asperger from the 1940s, who described a form of autism in children with normal or high IQs and normal or even precocious language development. Wing created the concept of a 'spectrum disorder', with Kanner's version of autism at one end – best represented by extremely impaired, often non-verbal children – and, at the other end, children with generally milder social difficulties and sometimes savant skills.

How autism grew after that is best understood by following it through the various editions of the DSM. It entered the DSM-3 in 1980 as 'infantile autism', with its main characteristics being gross distortions or deficits in language development, along with peculiar, sometimes rigid attachments to objects. To meet the diagnostic criteria, the abnormal behaviours had to be present before thirty months of age. In 1987, with the publication of the revised edition of the DSM-3, the diagnostic criteria were tweaked, with the need to have symptoms before thirty months replaced by the less specific statement, 'onset during infancy or early childhood'. That turned 'infantile autism' into the 'autism disorder', which immediately invited older children into the diagnostic group.

Over the 1980s and 90s, affiliated labels were added to account for people who did not meet the full diagnostic criteria for autism. The terms 'atypical autism' and 'pervasive development disorder, not otherwise specified' (PDD-NOS) were coined to describe social difficulties that could not be called autism because they were below the diagnostic threshold or unusual in some way. In 1994, with the publication of the DSM-4, Asperger's syndrome was added as its own discrete diagnosis, to describe children with normal or above average IQ and language development.

The DSM-5, published in 2013, moved the goalposts again. The symptoms of each subtype of autism, as described in the DSM-4, overlapped so much that it led to inconsistency in how doctors applied the categories. There was also a fairly arbitrary division of typical autistic difficulties into either impaired social interaction or impaired communication, each of which manifested as very similar symptoms in the child, adding further to the complexity of making a diagnosis. The DSM-5 aimed to address these issues by reducing the number of essential symptoms needed to qualify for the diagnosis. It also combined problems associated with communication and social impairment into a single list of symptoms. Subcategories like PDD-NOS and Asperger's syndrome were abolished, drawing them into a single diagnosis called 'autism spectrum disorder' (ASD), following on from Lorna Wing's work. It removed a prior requirement that the diagnosis be apparent before the age of three. It still required evidence of symptoms in the early developmental period – though no longer specifying an age, and, crucially, allowing for the possibility that they may be masked by learned strategies. Finally, it added sensory processing difficulties, such as an adverse response to specific sounds or textures, as a brand new feature.

In making the symptom list shorter, the DSM-5 made the diagnostic criteria for autism more specific, which might have reduced the number of people with autism.[6] However, the other changes invited older people and those with milder, atypical presentations into the group, so the overall effect was that the category grew. The DSM-5 also did something that reflects a common practice in medicine – it created a new category of diagnosis called 'social (pragmatic) communication disorder' to account for any people who might not now meet the new diagnostic criteria for autism. When a change in a disease definition risks depriving any person of a diagnosis, the solution is often to create a new label that does include them, so nobody will be left without. Diagnoses disappear with much less frequency than new ones appear. The DSM-1 had 106 diagnoses and the DSM-5 has nearly 300.

The consequences of these eight decades of gradual change have been dramatic for autism. Fifty years ago, the disorder was said to affect 4 in 10,000 people, but today the worldwide average prevalence is 1 in 100. In 2023 in California, 1 in 22 eight-year-olds were said to be autistic, while in Texas an estimated 1 in 64 were affected. In the same year, diagnostic rates in Northern Ireland were put at 1 in 20 children, compared with 1 in 70 in Australia, 1 in 134 in Belgium and 1 in 144 in France.[7 8 9]

There have been other shifts too. Autism used to be thought of as a disorder of boys. In the 1980s, the male-to-female ratio was 4:1. That ratio is now 3:1 and rapidly approaching 2:1.[10 11] There has also been a steady upturn in new diagnoses in adults, with one UK study showing a 150% increase between 2008 and 2016.[12]

These astonishing increases in the prevalence of the disorder are what has led to the concern for some that autism is now significantly overdiagnosed. Contributing to that concern is how difficult it has become to distinguish an autistic person from a non-autistic person. In 1943, autism was an infantile onset disorder diagnosed in children who had profound social communication problems. I see many people with this level of autism disability in my clinical practice. Most cannot communicate verbally. They require assistance for every ordinary activity of daily living, such as dressing and bathing. Most need constant supervision to keep them safe. Contrast that with today's growing autism community, which includes entrepreneur Elon Musk[13] and actor Anthony Hopkins.[14]

And yet, despite the substantial rise in diagnoses, many autism specialists are certain that there is still an *under*diagnosis problem, especially in girls and women. Their argument is that some in society don't think females can be autistic, so miss the diagnosis by not looking hard enough when presented with a girl or woman who may meet the diagnostic criteria. They say that females exhibit autistic traits in a slightly different way to males, meaning current diagnostic tests fail them because they depend too much on a male phenotype (observable traits). They also worry that autism is still missed because they see females as being particularly adept at masking the typical traits. In other words, some people are so good at 'pretending' to be 'normal' that teachers, parents and diagnosticians cannot see through the pretence.

Some of these underdiagnosis arguments have conceptual flaws. Take the theory that females are underdiagnosed because their autistic traits are too 'subtle' to be noticed. Autistic traits exist on a continuum, meaning we all have some, but in small amounts they do not cause impairments and therefore do not indicate a diagnosis of autism.[15] Lots of us are socially awkward, or become easily fixated on one thing, or misinterpret other's motives, or struggle with back-and-forth conversations. It is only when those aspects of our personalities are present with enough severity to cause impairment and exist in combination with other typical symptoms that they count as autism. It is worth asking, if scrutiny is required to pick up *subtle* symptoms, does that risk placing too much emphasis on social difficulties that are below the threshold of actually meeting diagnostic requirements? Is the impairment significant enough to qualify for a diagnosis, if such close scrutiny is required to notice the person's difficulties?

Another underdiagnosis argument theorises that girls are missed because they have a *different* phenotype (set of observable characteristics that, in this case, indicate autism) to boys. It is reasonable to surmise that some manifestations of autism will be at least a little different in girls. For example, the classical obsessive interests of boys with autism are trains and engines. A girl's obsessive interests might be different – experts in the field give examples like horses and celebrities.[16] An assessor who focuses too much on typical autistic male interests might easily dismiss a girl's pathologically all-consuming interest by asking the wrong questions. There tends to be more pressure to be social for girls, which could make autism harder to spot. A girl who is motivated to make friends might flit from group to group with nobody noticing that they do not have quality or lasting friendships.

The problem here is that some aspects of what is becoming 'female autism' are consistent with the original phenotype but some are not. If an autistic girl has a different obsessional interest to a boy, that still fits with the existing DSM description of the disorder because it is not the specifics of any interest that makes it part of the phenotype, but how utterly all-consuming it is. On the other hand, to say that an autistic girl is more socially involved than an autistic boy because girls are generally more socially motivated is at odds with the original phenotype. That described an autistic person as having a strong preference for 'aloneness'. Many diseases and illnesses present differently in males and females, so it is certainly good to scrutinise diagnostic criteria for this type of failing. However, changes to the fundamental description of what autism looks like have to be questioned. If a disorder is defined entirely by a behavioural phenotype and girls have an entirely different phenotype, then can you really say they have the disorder?

It is also important to interrogate the oft-made claim that we don't really know what autism looks like in girls because it has traditionally been thought of as a male disorder. Kanner's original description of autism featured the stories of eleven children – of whom three were girls: Vivian, Elaine and Virginia. The behaviours they exhibited were much the same as the eight boys that Kanner described. So examples of female autism have always existed and it is only lately that there has been a drive to create a new phenotype that will allow a diagnosis in people who do not fit with that description.[17]

Incorporating masking into the diagnosis also faces conceptual difficulties. Nobody is born knowing social rules. Children have to be taught how to behave within socially acceptable limits. So how can we distinguish between this and masking, which is also essentially learning social rules and employing them to fit in? The difference once lay in the amount of effort required to be 'normal'. A high functioning autistic person might be able to mask for short periods, however this is effortful and can't be maintained. The mask would inevitably drop to reveal the typical autistic traits. However, as the borders of what it means to have autism have moved, the assessors making the diagnosis no longer need the mask to drop. A person may appear entirely socially capable and it is enough for them to describe a great deal of effort to maintain that front for the diagnosis to be made. The masking theory tells us to assume autism is there even if it can't be seen.

Of course, while I might think that there are weaknesses in the underdiagnosis arguments, they are really only a continuation of what has been happening to the autism diagnosis, and to all diagnoses in the DSM, since the encyclopaedia's inception. The diagnostic criteria for autism have been steadily revised over eighty years and to change the phenotype further to pick up atypical presentations in women or milder cases is the most recent evolution. But diagnosis creep, the gradual expansion of diagnostic criteria, has to stop somewhere. There has to be a point at which a person who doesn't have the right symptoms and behaviours, or enough of the right symptoms and behaviours, simply doesn't have the diagnosis. But how can we know when that point has been reached? The answer should be simple in my view – when new people are being diagnosed but there is no evidence that it is benefitting them, or when the diagnosis is causing more harm than good.

The purpose of diagnosing autism is to give people support that will help their social functioning. Medication is usually only used in those with severe autism who have significant behavioural problems. For the mildly affected, treatment, if needed at all, takes the form of social and educational accommodations, and psychological and behavioural interventions. Since that kind of support might actually be helpful to any of us, even those who do not have autism, any concern about overdiagnosis may seem unnecessary. Except that no diagnosis, even one that doesn't lead to medication or unpleasant interventions, is risk free. There are potential drawbacks to both individuals and the group as the diagnosis concept expands. The diagnosis could disable individuals through the labelling effect, while over-inclusivity could make the diagnosis so broad that it loses all its power to prognosticate and point towards best treatment for the group.

A person with severe or moderate autism stands to gain a lot from treatment and faces minimal harms from the diagnosis, given their degree of disability and unequivocal need for support. It is the mildly affected who have the least to gain and who are more vulnerable to harm. The problem that this group is facing is that there has been a lot of enthusiasm for making more autism diagnoses but not a great deal of scrutiny of the potential harms. A 2020 study looked at 150 early autism intervention projects for evidence that they had considered the potential harms of the intervention alongside the benefits. Harms could include physical or psychological distress.

All the studies were assessing the efficacy of various behavioural interventions. These were not drug trials. Of the 150 projects examined, 139 did not measure harms.[18] When individuals withdrew from the trials, some groups did not even scrutinise why those children had withdrawn. Even those who withdrew for reasons that could be classed as adverse events related to the intervention were not reported as harms. Researchers seem to have been so certain that their interventions were always for the best, or at least neutral, that few seemed to feel the need to look for downsides.

Potential negative effects of a diagnosis are various. The stigma of autism has been associated with low self-esteem in children. Self-determination, the ability to be motivated and in control of one's own life, could be eroded by an autism diagnosis.[19] [20] Opportunities may be closed off to people with autism. For example, one Turkish study suggested that prejudice, a child's awareness they are different and time spent in special education meant autistic children were less likely to be included in sports.[21] The impact of the assessment process, medicalisation and parental stress have an untold effect. The diagnosis might become a self-fulfilling prophecy as some will take the diagnosis to mean they can't do certain things, so won't even try.[22] And why would they, with a neurodevelopmental problem presented as an immutable scientific fact? Eric Fombonne, a child psychiatrist with a special interest in autism, has spoken of his concern that autism is overdiagnosed, expressing the belief that 'carrying an ASD diagnosis may unduly constrain an individual's range of social and educational experiences and have long-lasting effects on their identity formation.'[23]

An autism diagnosis can also obscure psychosocial difficulties that really need to be addressed. When children who are bullied, have eating disorders, who have been abused, who self-harm or who are suicidal have these experiences explained through the lens of autism, it risks blaming their personalities and their vulnerability on something that is out of their control and cannot be fixed. A diagnosis that attributes all psychosocial problems to faulty wiring of a child's brain makes it too easy to stop people asking: what in this child's life is making them sick?

There is evidence that early intervention, especially in preschool years, can help children develop socially, but in adults there is no good evidence for either lack of harm *or* benefit with the diagnosis. Anecdotally, many adults report improved self-acceptance when they learn they have autism, but there is no proof that translates to any more tangible advantage, like an easier progression through life, better relationships or success in work, or other ambitions. The drive to recognise mild autism in adults has taken place before anyone has measured the trade-off between feeling validated and the potential impact of negative perceptions. People with autism can be perceived as less capable, by themselves and others, which might narrow their possibilities instead of widening them. The decision to tell lots of people that their brains are 'different' or neurodevelopmentally 'abnormal' has happened before anyone has established what happens to a person's motivation when they are told something like that.

Overdiagnosis is not misdiagnosis. It is very hard to spot overdiagnosis at an individual level because, for the most part, people are relieved to get a diagnosis if it explains symptoms or difficulties. Overdiagnosis has to be assessed at a population level. You can spot it when an increasing number of people have a diagnosis but there is no corresponding improvement in long-term health measures for the population. The theory is that mild autism diagnosed in a timely manner will result in support that will help a person to prosper. Rates of autism diagnoses have been steadily rising for over thirty years – time enough to see the benefit of these diagnoses – and yet it is hard to find any.

In the US, the CDC shows rising rates of depression and anxiety in children, increasing from 5.4% in 2003 to 8.4% in 2012.[24] In the UK, the proportion of the population reported having ever self-harmed increased from 2.4% in 2000 to 6.4% in 2014. Eating disorders in teenage girls rose from 0.9% in 2017 to 4.3% in 2023. Despite educational accommodations, school dropout rates in the UK, Australia and many European countries are at an all-time high. Mental health problems are rising in children and adults of all ages. In the UK, there is a rising number of people off work due to health conditions and more than 50% of those have a mental health problem.[25] A third of the 16–34-year-olds who are off work in the UK due to illness cite depression and anxiety as the reason.

Of course, there are many other factors that impact mental health but, still, if the increased number of children with an autism diagnosis was indeed just a correction of an under-diagnosis problem, one would hope to see an improvement in mental health or social functioning in some group by now, in some measure, but the opposite is true. There are more people with autism *and* more people with other mental health conditions too. One in five children in the UK has a mental health disorder and this figure is gradually rising.[26]

In writing this book I spoke to a psychologist who is involved in diagnosing autism and in developing new phenotypes to allow more diagnosis in females. I asked about the possibility of overdiagnosis and she replied, 'We are diagnosing people to help them in their life. That's how we reconcile that there is a need for a diagnosis.'

A diagnosis may be well-meaning and it may provide a feeling of relief, but that cannot justify overdiagnosis if the harms have never been properly considered and the long-term outcomes have not been sufficiently examined. Autism is a diagnosis either sought out or accepted by people who are struggling and in need of help, that is certain. In the absence of other sources of care, a diagnosis can potentially provide it. If it helps people to advance in their life or gives them lasting relief, then the diagnosis is probably worth it. But there is no proof to say that it does, or that the feeling of validation turns into something more meaningful. The struggles are real but medicalising them may not be the solution.

*

'What autistic people want is an affirmative approach to diagnosis,' Miles advised me. 'Ask us what we can do, not what we can't do. The medical tests are too negative. They leave autistic people feeling judged and demeaned.'

Miles is sixty now and was diagnosed as autistic in his mid-fifties. He's a retired banker, married with three adult children. Two of those, a daughter and a son, were each diagnosed with autism in their thirties. His son was the one who suggested Miles should also be assessed. Miles had troubled relationships and a chequered history in banking. The longest he'd stayed working for any single business was three years. Jobs tended to come to an end after conflict with colleagues or clients. Miles's career never progressed as he thought it should have. So Miles took an online assessment and it seemed to confirm his son's suspicion. Afterwards, Miles met a psychiatrist in the private sector, but that doctor determined that Miles did not have autism. Unhappy, Miles sought a second opinion. The next psychologist agreed with Miles that he was autistic.

With the diagnosis confirmed, Miles saw his whole life in a different light. He realised that, as a natural loner who could not work in teams or tolerate a noisy environment, he had always been doomed to fail in banking. He took redundancy shortly after. These days, aside from occasional consultancy work which he does from home, Miles spends most of his time gardening.

At the start of our conversation, I asked Miles how he wanted me to refer to his diagnosis.

'Don't call it a disorder,' he told me.

I warned him that although I would not refer to Miles as having a disorder, I expected to use the term autism spectrum disorder quite a lot, as that is the official DSM terminology.

'And why is it even in the DSM?' Miles asked. 'That's another error. It's not a mental health problem.'

Miles rejects most of what the DSM has to say about autism. He doesn't accept the official label 'autism spectrum disorder' since he regards both the words spectrum and disorder as pejorative. To him, autism is a difference. He does not acknowledge autism as a disability; although, he admitted to using autism as the basis for medical retirement. Miles's views are not niche. Even the UK's National Autistic Society has replaced the word 'disorder' with 'condition' on its website.

'Would you accept that some people with autism are much more disabled than others and that, for them, the concept of an affirmative diagnosis doesn't work so well?' I asked. I had in mind my own patients, most of whom are far too disabled by autism to live independently.

'No. I wouldn't accept that,' Miles said. 'Who says those people have learning impairment? You? Autistic people are underestimated all the time. Just because someone is non-speaking, doesn't mean they are not intelligent.'

I asked Miles what his diagnostic journey entailed. He described an assessment that took place one-to-one over ninety minutes. He attended alone. At it, he had described the difficulties he'd had both as an adult and in childhood. He told them he was an unsociable child. He used to pray for rain so he didn't have to play in the playground at lunchtime. He hated school and thought of it as a lonely place.

'Why do you think the diagnosis of autism was so slow to come to light?' I asked.

'I'm very high masking. I have faked being neurotypical for years.'

Miles told me he has never unmasked in public. It was something he only does in the safety of his own home.

'What would you consider to be your main autistic characteristics?' I asked.

'I'm guessing you want to know what's *wrong* with me,' he said, sighing with a sort of weariness at the predictability of my question. 'This is what the community is tired of. Autism is not just about negative traits. Some of the cleverest, most talented people I know are autistic.'

He was correct, of course. Being autistic does not mean a person cannot be intelligent, creative and able, and he was also right to say that I was still trying to understand what was *wrong* with him. My question was supposed to elicit something of the impairment that qualifies this as a medical problem.

'Autistic people don't want to be fixed,' Miles corrected me. 'Neurotypicals love to think autistic people are the problem but making us fit into your world is our real problem.'

Miles is one of a very large and growing number of autistic people who disagree with the medical profession's deficit-based approach to diagnosis, preferring an approach that concentrates more on what makes an autistic person special. They advocate for encouraging autistic people to be their most authentic autistic self. For people with this view, the concept of treatment is offensive. To suggest that a person with autism should suppress any of their autistic traits would be like suggesting conversion therapy for homosexuality.

*

There seems to be a growing public perception that the reason there are so many people with autism these days is because it is easy to get an autism diagnosis. But actually, despite the lack of biological markers, the diagnostic process is detailed and robust when done properly. In most parts of the world, an official diagnosis of autism is made using two semi-structured interviews called the Autism Diagnostic Observation Schedule (ADOS) and Autism Diagnostic Interview-Revised (ADI-R). Referred to as diagnostic tools, the ADOS assesses current communication and behavioural difficulties, while the ADI-R focuses on early development with particular attention paid to the period between the ages of four and five. Ideally, each interview should be carried out by a different expert practitioner and can take several hours to complete. Multiple sources of information should be taken into account, including observations of the person in different environments and, in children, corroborating evidence from teachers. An optimum assessment is videoed so that it can be reviewed in a multidisciplinary setting by a whole range of experts who can come to a consensus diagnosis. But either way, a diagnosis should be made by more than just a single clinician.

Just like the tests for Lyme disease, the ADOS and ADI-R come with many caveats. An assessment for autism should never be carried out during a crisis period in a person's life and certainly not when a person is suicidal. The diagnostic tools are not supposed be able to deliver a diagnosis by themselves, without other input. Like Lyme disease, like most medical conditions, it requires a clinical context – the patient's story. A person might have communication difficulties for any number of reasons, such as a learning disability or anxiety disorder, so an assessor has to have a broad medical training to ensure they have the experience to recognise other non-autism related causes of communication problems.

Certain aspects of the assessment present regular challenges to the assessors. In adults, it can be very difficult to establish that there were problems in early development. There is no school to provide corroborating stories. Adults attending without parents can leave the assessor with very little to go on to fulfil this essential requirement for the diagnosis. Impairment, also essential for the diagnosis, can also be difficult to quantify. The DSM does not define what counts as impairment at the mild end of the spectrum.

An expert assessor told me, 'There is no clear answer to that because what is impairing to one person may not be for another. If a person tells you that the lights in the supermarket are very hard to tolerate, I then ask if they can actually go into the supermarket. If they can, but they're exhausted afterwards and have to go to bed – then that's impairment. They can do the thing but it comes at a cost. Even their own family may not see that sort of impairment because they remove themselves, needing time to recalibrate.'

It is entirely at the discretion of the assessor to decide whether tiredness after the supermarket that goes unnoticed by a person's family really counts as impairment. One might decide yes, that is enough, and another might decide no. The subjectivity of this measure leaves the door wide open to a broad range of different clinical practices.

The concept of masking has also had an undermining effect on the diagnostic tools. If a person's assessment does not meet the score needed to diagnose autism, it can be overruled by their description of their subjective experience by saying that their autistic features are masked. Which means a person can get a diagnosis of autism even if they exhibit minimal typical autistic behaviours during the assessment. Masking allows people a diagnosis of autism even if the ADI-R and ADOS do not detect it.

There is also uncertainty about the reliability of these diagnostic tools in different ages and ethnic groups. The tests were developed by looking at validation samples. These samples are groups of people diagnosed with absolutely typical autism by a group of experts, so a good place to begin to test accuracy. But most of the people in those validation groups were white, with English as their first language. There is no proof these tests are equally applicable to older adults, ethnic minorities, non-English speakers, people with physical disabilities and those with complex psychiatric problems. But they are used as if they are.

Speaking to an autism assessor, I was impressed by the level of detail and number of people involved in making a diagnosis. But that complexity also presents a problem both to the quality of diagnosis and to the science of autism. Pressure on healthcare systems to make more diagnoses, more quickly, is likely contributing to the overdiagnosis and misdiagnosis of autism. There is not always enough time or money for assessors to uphold the standard of testing that the diagnostic tools require. There are currently 1.2 million people waiting to be assessed for autism in England[27] and the average wait is one year. It is difficult to continue to provide lengthy assessments that involve large teams of people with waiting times like this. In the US, to facilitate easy access to assessments for all, some autism centre teams have had requests to train education professionals and other primary care providers in the ADOS. In the state of Oregon, a child can already be diagnosed by their school without a medical evaluation. But the ADOS was only ever intended to supplement a *clinical* diagnosis. A teacher may be incredibly informed about children, development and education, but they are not diagnosticians. They do not have the wide knowledge of psychology needed to spot other medical disorders that could easily be mistaken for autism.

The consequences of the pressure on diagnosticians are being felt. One 2022 US research project recruited children with a diagnosis of autism that had been made in the community. As part of recruitment, the children were reassessed for autism by research standards to confirm the diagnosis. Of those examined, 47% did not actually fulfil the research criteria for autism, suggesting they were overdiagnosed or misdiagnosed at their community assessment.[28] A recent UK study found that diagnostic rates varied enormously between autism assessment centres, with some facilities making a diagnosis of autism in 85% of clients and others in only 35%.[29] This can only mean that, as robust as the diagnostic tools may be when used correctly, they are not being used in the same way by all practitioners.

Perhaps most sobering of all are the words of Allen Frances, retired professor of psychiatry, who was chair of the task force that developed the criteria for autism in the DSM-4. He has openly expressed regret for his role in making the diagnosis so inclusive and has cast doubt on the quality of testing. 'Because the diagnosis of autism is so consequential and so frequently carelessly done, parents and adult patients should always get a second opinion whenever possible,' he advised. He has also expressed concern about the negative effects of overdiagnosis, saying 'an inaccurate diagnosis can cause harmful stigma, hopelessness, reduced expectations and misdirected treatment.'[30]

Varying approaches to assessments are also having an impact on research standards. Researchers looking to recruit people with autism for various studies into the disorder don't always have the budget or time to carry out a lengthy, expensive, formal assessment to ratify the diagnosis in those they are studying. Therefore, some have started to allow basic screening questionnaires and self-diagnosed patients into their research. Basic screening tools are only supposed to whittle a large population down to a smaller one, which will then be subjected to more rigorous testing. They have a very high false positive rate. And in self-diagnosing, many people distil autism down to a small group of traits, such as difficulty with friendships or social anxiety or noise intolerance, not understanding the complexity of clinical diagnosis and all that is required. Inviting self-diagnosed people into research is contrary to scientific principles and will return very inaccurate results that are impossible to interpret.

Also contributing to overdiagnosis and misdiagnosis is the pressure on psychologists, doctors, teachers and parents to give struggling children a medical label. Parents are worried about their children and perceive this as a way to help them. For many, a medical diagnosis is the only way to access extra support, both medically, financially and as school accommodations. Insurance plans in the US will cover behavioural therapies for autism, saving families tens of thousands of dollars per year, but only with an official diagnosis. For schools, a diagnosis can sometimes justify higher staffing levels and extra time in exams for the child. If a child is supported to perform better that is good for the child and reflects well on the school. So much so that some parents have found it difficult to resist the diagnosis when they might have preferred their child to be supported without a label. In researching this book, I met several parents who felt under intense pressure to have their child assessed for autism and who had fought hard to resist it.

It is also hard to ignore the social contagion element that drives overdiagnosis and which is changing the meaning of what it is to have autism. Social media is filled with people celebrating an autism diagnosis – many of whom misrepresent the disorder, giving a skewed idea of what the diagnosis is.

One 2023 study found that #Autism had attracted 11.5 billion views on TikTok. But of the top 133 most-viewed videos, only 27% contained accurate information.[31] The respected mainstream media in all forms is also guilty of creating a somewhat romanticised view of the disorder, presenting it as something that makes a person special rather than a developmental disorder that causes impairment.[32] Remote diagnoses of autism are given to successful people with covetable careers and lives, even if those people have never publicly claimed to have a diagnosis, with philanthropist Bill Gates[33] and director Tim Burton[34] among them. Deceased figures have also been diagnosed in retrospect – Michelangelo, Charles Darwin, James Joyce and Albert Einstein.[35] Celebrities make bold statements about neurodevelopment disorders without seeming to understand much about them. In 2018, singer Robbie Williams told a newspaper, 'There's something missing in me, I have big blind spots. Maybe Asperger's or autism. I don't know what spectrum I'm on – I'm on something.'[36] The general public has also developed a very casual way of referring to people with mild social awkwardness as 'on the spectrum'.

Uta Frith is a retired professor of cognitive development and a highly respected pioneer of autism research whose work is still very influential in the field. She is of the view that this blurring of the line between those with autism and the rest of the population could muddy any attempt to understand the condition. As people with autism move from being a narrowly defined and homogeneous population towards an inclusive and heterogeneous population, it will become increasingly difficult for researchers to determine why autism occurs and how to support those affected. In Frith's words, 'the diagnosis of autism has been stretched to breaking point and has outgrown its purpose. If the purpose is to predict what an individual's needs are, this is no longer possible.' So, if those with pronounced autistic traits or who are even non-verbal with severe autism are studied alongside those whose symptoms are so minimal that they were not noticed until late in life, it becomes less likely that this group will have enough in common to allow any trial to be able to spot a single cause or find a treatment that will work for all. Frith has advised, 'researchers need to think hard about how to disentangle the underlying conditions in individuals now all labelled autistic. Without such an effort, research into the causes of autism will become meaningless.'[37] It does not help that there is considerable social pressure from some people with mild autism to dispense with terms like 'severe' autism and pressure on scientists to consider all people with autism as a single group. That limits scientists' scope for drawing more similar subgroups of people with autism together to improve the chances of finding a commonality between them.

The conversation around long Covid was shaped by social pressure, probably to its detriment, and it is likely that is also happening in autism research. In 2021, autism researcher Simon Baron-Cohen launched a genetic study called Spectrum 10K that aimed to look for potential genetic causes of autism. The study promised to be the biggest of its type in the autism field. But, a few weeks in, the study was stopped as a result of public pressure.[38] Autistic people expressed concern that it had gone ahead without meaningful consultation with the autism community. Campaigners were worried that it threatened people's privacy and that it could be the start of a programme of eugenics. As we will see in later chapters, population-based genetics studies like this are very common, with many producing extremely useful results that certainly benefit patients – although that is not to say they are without their problems, which will also be discussed. Baron-Cohen's study is unique in being the only one to attract any such objection. Those objections did not come from individuals with severe autism or their families, but from the mild and masked group, many of whom, like Miles, also object to the use of the term 'severe'.

That cuts to the heart of my personal concern about the over-diagnosis trend in autism, which is the effect that overdiagnosis seems to be having on Kanner's autistic children and on the adults with severe autism like those I look after. Those severely disabled people who could not live in the world without support. They are now in a queue for resources alongside people with considerably fewer difficulties. My patients with severe autism are not represented on TikTok, where people celebrate their great relief at being diagnosed and call for affirmative diagnosis and ask for words like 'spectrum', 'disorder' and 'impairment' to be removed from the language of autism. Those with the greatest need are becoming invisible.

*

'Looking back, it was obvious,' Agatha said. 'I don't know how we missed it.' She was telling me about her son, Elijah, now aged twenty. 'When he was one, he had no words. He flapped his hands continuously. He was quiet and I remember thinking we were lucky to have such a good baby. When he was a bit older, we could just pop *Peter Pan 2* on and he'd sit and watch it for hours. And he'd jump up and down and say eee eee eee, over and over and over . . . and we used to think, *oh, isn't that sweet*.'

Elijah has an older sister who has high functioning autism. But even with that experience in the family, Elijah's parents didn't spot the early signs. When Elijah didn't talk, Agatha told herself that boys learned to speak later than girls. Plus, Elijah was affectionate and cuddly.

'He wasn't a loner, which is one of those tropes about autism,' Agatha said. 'His grandmother did say she found him hard to bond with, but I didn't because I suppose I loved him so much. I think I missed the signs because I didn't really want to know.'

When Elijah turned two, he still couldn't speak. His parents took him to a clinical psychologist and the first thing they asked was, does he point? He didn't. Agatha learned that pointing was an imaginative gesture and that not pointing was an early sign of autism. His other behaviours fell into place. Elijah insisted on lining up his toys, reflecting his need for order. His constant hand flapping and bouncing were consistent with repetitive behaviours, or stims. After a detailed assessment, Elijah was diagnosed with a learning disability and autism.

As he grew older, Elijah's autism became even more apparent. Routines were very important. He would only eat one meal – chicken nuggets and beans. If, after finishing his dinner, any ketchup remained on the plate, it had to washed immediately, even if they were eating out in a café. Although a very loving child, he started to exhibit many of the challenging behaviours seen in people with severe autism. He hit his head on the floor repeatedly and punched his own head if upset. He ran up and down manically, irrespective of where he was. He had no sense of risk or danger. Agatha had to watch him constantly to keep him and other people safe. If he was frustrated, he could bite and hit out at strangers. He needed round-the-clock care and help with every sort of activity.

'He was so beautiful,' Agatha said, 'He had lovely curly blond hair, but when he sat on people's knees I'd have to warn them to be careful in case he suddenly head-butted them.'

For Elijah at twenty, that hasn't changed. He still bounces for hours at a time and hits out at strangers as he did when he was two years old – except now he is a six-foot-tall, fifteen-and-a-half-stone man.

'What frustrates me,' Agatha said, 'is that the whole autism discourse is being dominated by the *mild voice*. And that is having real-world consequences for my child. It affects services and funding. A school for autism opened up nearby but they'll only take children with a normal IQ who can meet the national curriculum. Increasingly, schools are only catering for the high functioning. Hospital waiting lists are too long. Children like Elijah are being neglected.'

Elijah attended mainstream school in the first instance. He didn't make friends. He didn't seem to know what friends were for. Still, mainstream school helped him accommodate to crowds. It also helped the other kids learn to be around people like him. He had a one-to-one tutor and lessons were differentiated for him. If the class was doing maths, Elijah did a counting task. He learned to count to a basic level and learned to read phonetically. However, by age seven, the gap between Elijah and the other children had grown too big, so his family moved him to a special needs school.

Aged twenty, Elijah can speak in short, simple sentences of up to five words. When I met him, he was gripping four fluorescent-coloured, highly textured plastic caterpillars close to his chest. If Elijah is awake, he has two things in his hands – his caterpillars and his iPad. He also loves swimming, tennis and discos. He still likes bouncing, so has a trampoline. He enjoys going to Starbucks for a latte.

Elijah continues to learn basic life skills. He has a job, filling fruit bowls and emptying the dishwasher at a local business. He can go to the shop and make small purchases. All of this is supervised. Left alone, Elijah would run in front of traffic. He requires the same level of care as a small child.

'He's learned to use his own Amazon account,' Agatha smiled. 'He buys caterpillars online. After he's bought them, he goes to the window and says "postman, postman, postman", over and over until they arrive.'

Everything that Agatha described is consistent with classical autism. His lack of interest in friends. His routines. His meltdowns.

His obsessions with very specific objects. Elijah listens to the same song or YouTube video so continuously that Agatha has to persuade him to stop, lest the whole family go mad.

The family have worked very hard to help Elijah control his aggressive outbursts so he can live as normally as possible in the world. They use techniques to calm him if he gets upset in public. Agatha's worst fear is that he might be institutionalised if he cannot control his behaviour.

Agatha and many parents of children with severe autism have grown concerned about how the current conversation around autism misrepresents the disorder and sidelines their children.

'It's a neurological disorder from birth. The public think they know what autism is by listening to the voices represented in the media. But these are people with a normal IQ, who are verbal, who have proper jobs, who can attend meetings or log onto Zoom. These mild and self-identified people have no concept of people like Elijah. They assume all autistic people are a bit like them, but Elijah is not like them except for some very small commonalities.'

Agatha had voiced exactly what had concerned me as I interviewed people for this book and also watched autistic people telling their stories to the media. One question was constantly on my mind as I spoke to Miles, and even Poppy. They speak for people with severe classical autism but have they ever met one? As things stand, autism representation in the public arena, on the boards of businesses, charities and academic panels, is heavily skewed towards those with lesser care needs, most of whom live normally in society without any obvious outward autistic traits because they are able to mask.

Some autism activists criticise parents of non-verbal autistic children, like Agatha, for speaking for their children. They believe only autistic people can speak for autistic people. Strategies to help people like Elijah to control their behaviours have been called cruel on the basis that an autistic person should never be asked to change their autistic traits because that is like asking them to change their identity. Miles was certainly of the opinion that a person with autism should be encouraged to display all their autistic traits fully and was completely against any medical intervention that had the opposite effect.

'I needed to teach Elijah to tolerate sounds he couldn't cope with, like babies crying. If I didn't teach him that, I'd be narrowing his world. If I took the advice of people with mild autism on Twitter and on the radio, Elijah would end up in residential care with risperidone being pumped into his veins and five people holding him down,' Agatha said. Risperidone is a strong antipsychotic drug used to control aggressive behaviour in autism.

To keep Elijah safe, Agatha tells people he has the mental age of a four-year-old. She wants anybody caring for him to remember that he might look like a man but he still has the needs of a child. But in referring to her son in that way, she has been accused of infantilising him.

'The way the mild and self-identified autism movement talks about autism feels to me as if they dislike intellectual disability. People don't want to believe that their kid has a low IQ or is below average. They want to believe that inside is a secret smart kid. But I love Elijah for Elijah, with his IQ below 50. You and I learned to read easily but for Elijah, every word he learned was so hard won and so precious. I value him for his actual intellect not for some fantasy.'

Elijah has very high care needs and every bit of progress he has made has been fought for by him and his family. In my day job as a neurologist, I meet lots of people like Elijah. I see many parents like Agatha interact with their severely disabled adult children and it is often awe-inspiring. They have come to know their non-verbal children in the most intimate way, until they can read their needs from small changes in expression and temperament.

'The public don't really recognise Elijah's part of the spectrum as autism anymore,' Agatha told me. 'I have protocols for his meltdowns. I create distance so he can't head-butt me. I have to say very sternly, "You need to calm down." I say it over and over. I have to show him that this is serious so he knows he has to regain control. I was doing the protocol on the bus recently and I realised all the other passengers were staring at me. I had an audience,' she smiled. 'So I turned around and I said, "*This* is what autism looks like, not what you saw on the telly last night!"'

'What did they say to that?' I asked.

'They looked away.'

4
The Cancer Gene

I don't think that many people will be surprised to find disorders like autism, long Covid and chronic Lyme disease in a book that is focused on overdiagnosis and on the power of a label to both validate our suffering and, sometimes, to exacerbate it. Certain aspects of medicine and health have grey areas that everybody recognises. Mental health disorders will always be at least a little controversial because there are no objective measures to say where 'normal' ends and 'abnormal' begins.

However, some of the biggest overdiagnosis and overmedicalisation controversies do not lie in the field of psychiatry, as you might think, but in the high-tech world of cancer and genetic diagnosis. One might expect that predictive tests for cancer are ones most of us would want to take if offered. And very soon you might receive that offer. I described the learnings from the predictive tests for Huntington's disease in chapter one – the next stage is to give more people a chance of an advanced diagnosis for a much wider range of conditions.

If and when that offer comes, I would suggest you keep in mind all the lessons so far. That diagnosis is first and foremost a clinical art. That the line between good health, illness and disease is often blurred. That a diagnosis can be right but treating it can still do more harm than good. Because, as we'll see, these rules apply even to the most cutting-edge diagnosis.

*

The story of breast cancer has changed a great deal in the forty years since I entered medical school. Back then, it all too often felt like an inevitable death sentence. I have a particularly vivid memory of a young woman with end-stage breast cancer who I met on a surgical ward when I was quite newly qualified. She was an architect in her early thirties who everybody was desperate to save. She was due to get married and had a sparkling diamond engagement ring that her fiancé used to bring with him when he visited so she could wear it for a while. She had lost so much weight that it had become far too big for her. He'd take it home with him at the end of the visit to keep it safe. Her cancer had spread by the time she was diagnosed. I was in my twenties and it was chilling to see a person so close in age to me, with the sort of life I would like to have, so diminished.

Treatment for advanced breast cancer in the 1990s was not as successful as it is now. All the surgeries and chemotherapy and radiotherapy she endured had no realistic chance of saving her. If she was older, I suppose her treatment might have been less aggressive. She would have been allowed to die at home or in a hospice or anywhere nicer than the side room of a general hospital ward, but she never even had the chance to leave the hospital after her first admission. She was so young and otherwise full of hope that everybody – her, her partner, her family and her doctors – were desperate to eke out as much life as they could from this impossible situation. She was given chemotherapy almost up to her last day and got married in the hospital shortly before she died. Lots of staff referred to the wedding as romantic, but for me it was too poignant to tolerate.

That young woman had a family history of breast cancer. Her mother had died of it in her forties. Familial cancers were well recognised by then, so she knew she was potentially at risk long before she was diagnosed, but there wasn't much she could do to prevent it. She died in 1992. Two years later, there was a scientific breakthrough which could have saved her life if it had happened just a few years earlier.

Chromosome 17 had been identified as the likely site of a breast cancer gene in 1990 but its exact location was unknown. The BRCA1 gene was discovered on chromosome 17 in 1994. BRCA2 was found on chromosome 13 a year later. People often think of BRCA1 and 2 as 'cancer genes', but actually they are the opposite. Every person carries these two genes and, when healthy, they promote the repair of DNA and suppress the development of tumours. However, a variant (or mutation) in one of these genes can mean they lose their protective effect. There are numerous gene variants that increase the risk of various cancers, but the best known of those are the BRCA variants. To date, more than 4,900 high-risk cancer variants have been found in BRCA1 and BRCA2.[1] The increased risk of cancer applies to breast, ovarian, fallopian tube and peritoneal cancer, and, to a lesser extent, prostate and pancreatic cancer, and melanoma.

Like the Huntington's disease (HD) gene, BRCA variants are used to make a predictive diagnosis in healthy people before the disease has had time to develop. However, the predictive value of a BRCA variant and that of the HD gene are not the same.[2] A person with a family history of HD who is found to have the gene is guaranteed to develop the disease if they live long enough. Whereas a BRCA variant is only a risk factor. It increases the likelihood of certain cancers very significantly but still, not everybody who carries a cancer variant will get cancer. Different variants in BRCA come with a different degree of risk. Some are more deadly than others. Decades of experience of cancer genes has allowed geneticists to create risk prediction models for each of the thousands of variants. Approximately 12% of all women will get breast cancer in their lifetime. That increases to between 60% and 85% in women with certain BRCA1 variants and to between 40% and 65% in women with variants in BRCA2. Two per cent of women will get ovarian cancer in their lifetime. With high-risk variants in BRCA1 and BRCA2, that increases to 39–58% and 13–29% respectively.[3]

Another vital difference between HD and cancer is that HD is incurable. Women known to carry a high-risk variant of BRCA have the option of taking action to prevent cancer. The foremost of those is risk-reducing surgery. That means having their healthy breasts, ovaries and fallopian tubes removed before cancer has a chance to take hold.

*

Roisin was only eight years old when her grandmother was diagnosed with ovarian cancer and her mother with breast cancer. She had to watch both women undergo chemotherapy at the same time. It was touch and go for Roisin's mother, who was only in her thirties at the time. She spent a week in a medically induced coma and the family thought they would lose her. She survived but Roisin's grandmother didn't. She died in her fifties, four years after her diagnosis.

That was only the beginning of Roisin's family's journey with cancer. Twelve years later, when Roisin was twenty and pregnant with her first child, her mother sat her down in the family kitchen and, over a Chinese takeaway, announced that she had stage three ovarian cancer. It was the same cancer that had killed her own mother. Thankfully, Roisin's mother overcame that cancer too, only to develop breast cancer for the second time a few years later. She survived again. She is now in her fifties, happy and healthy. Roisin and her mother are each other's best friend. They have been through a lot together.

Only 5–10% of breast cancers and 5–15% of ovarian cancers have a genetic cause. However, in families like Roisin's, where there has been both breast and ovarian cancer and cancer has been diagnosed before the age of forty, the likelihood of a genetic cause is very high and genetic testing is recommended. Predictive genetic diagnosis in healthy people is not recommended below the age of eighteen. Roisin waited until she was twenty-five to be tested. While HD predictive diagnosis usually requires a bare minimum of three counselling sessions in the UK and two in the US, and is often quite drawn out, cancer is preventable and treatable so the road to testing is much quicker. Which is why Roisin met a genetic counsellor only once. At the end of that single consultation, Roisin gave a blood sample and went back to her life. A few short weeks later, she was at work when her phone rang with a number she didn't recognise. That was how the counsellor told her that she had a high-risk BRCA1 variant that gave her an 87% chance of developing breast cancer and 60% chance of developing ovarian cancer.

I met Roisin ten years later.

'Looking back, it was all too quick,' she told me.

Roisin partially expected the result but it was all still very sudden. The family were devastated. Roisin's mum more so than anyone else because she knew she was the one who had passed the gene to her daughter. Both women were immediately terrified that Roisin would get cancer. Overnight,

Roisin went from being perfectly healthy to always looking out for cancer. That was why, eight months after the positive result, aged twenty-six, Roisin had a double mastectomy and breast reconstruction. It was an instant decision followed by a very difficult aftermath. Everything about the experience of surgery was traumatic. Roisin didn't feel that her surgeon understood the magnitude of having her healthy breasts removed. The medical care to treat Roisin's physical wounds was excellent but very little was offered to address what she was going through psychologically. Which is perhaps why, eighteen months later, when Roisin became pregnant with her second child, the consequences of her decision really hit home for the first time.

'I went to all the classes and it was all *breast is best!* And it made me feel so lonely. They kept telling us that's how you bond with your child, so how was I supposed to bond with my child?'

Roisin struggled with other people's reactions. An aunt, referring to the mastectomy, said, 'Why would you do something like that to yourself?' Others called her brave. That rankled because in her eyes she hadn't done something brave, she had done something drastic because she had no choice. Not that she regretted the surgery. She had seen her mother fight cancer three times and win. She had also seen her grandmother die. Her decision to have surgery was made to save her own life, but living with the decision was still incredibly difficult. It affected her relationships. Things her partner said made it seem he was no longer attracted to her. Shortly after her second daughter was born, Roisin left him. She was terrified ending that relationship, worried that no man would ever find her attractive again. She wasn't happy with her new body. She later had the breast reconstruction revised, though the surgeon was reluctant as they didn't think there was anything wrong with the first reconstruction. Roisin had to push hard for further surgery.

'How did you feel about your body before the surgery?' I asked.

'I had great boobs! I loved my boobs. I remember waking from surgery and wondering what the hell I'd done. Breastfeeding went terribly with my first child and I'd always thought I'd do better the second time, but then I couldn't. I hated my new breasts. They looked stuck on.'

Roisin doesn't really like her revised breast construction either. They feel unnatural to her and she has no nipples. Some women choose to keep their nipples and areola for cosmetic reasons but removing them gives more protection against cancer, which was Roisin's priority.

'I'm working hard to love them,' she told me.

Roisin waited until she was thirty to have her ovaries, uterus, fallopian tubes and cervix removed. That was even tougher. She knew that the surgery meant she would never have more children but, once again, it wasn't until after the surgery that the reality of that sunk in.

'I don't think I would have had more kids, but who's to know,' she said.

For the second time, the care she received concentrated on all the physical aspects of surgery and on cancer prevention, but little was said to her about the inevitable psychological impact of her choice. This time, she sought the help of a psychologist. She also turned to a menopause support group at work. She was the youngest in the room by twenty years. They were welcoming but she didn't feel welcome. Or perhaps she felt she didn't belong.

'I thought my overnight menopause would be difficult for women who had a natural menopause to understand.'

A surgical menopause is much more abrupt than normal menopause. A bodily change that would normally take place gradually over several years, allowing time for a woman to adjust, happens after one surgery. Roisin gave her surgeon an HRT patch before she went into surgery and asked him to attach it as soon as the job was done. But it didn't stave off every menopause symptom.

'I'm an empty vessel now!' Roisin laughed. 'I've had everything removed.'

'But you feel you made the right choice with both surgeries?' I asked.

Roisin thought for a long time before answering. 'I did it out of fear. I've been around cancer my whole life. My mum has nearly died more than once and she was so scared for me. I felt pressure to have the surgeries. I'm still not really sure if it was my decision. But did I do the right thing?' She sat back and contemplated again. 'Yes, I'm glad I did it because the constant fear I will get cancer is gone . . . Well, it's not completely gone . . .' She smiled. 'Not all the breast tissue is gone so I could still get breast cancer and I recently read that BRCA can cause pancreatic cancer and you can't have your pancreas removed . . .'

Mastectomy and salpingo-oophorectomy (removal of the ovaries and fallopian tubes) reduces the risk of the respective cancers by 95%.[4] If she had the choice to make again, Roisin knows she would still have the surgeries. She is less scared of cancer now and her life has moved on in positive ways. She met a new partner and is due to get married soon. He loves her body, she told me. She often makes jokes about the surgery but he understands that those jokes are advising him to be careful with her, which he always is. She has two wonderful daughters. They might also have the cancer variant but are far too young to be tested. Roisin isn't worried for them because she knows she can support them through the decision to have surgery, although hopefully medicine will have moved on by then in ways that mean it won't be necessary. Her mother, now fifty-eight, has survived three bouts of cancer and the pair remain each other's main support. Her father is always there for her too, a quiet Scottish man who has seen his wife and daughter go through a great deal. Roisin's brother also inherited the high-risk BRCA1 variant. It has increased his risk of breast cancer from 0.1% to 1% and also increases his risk of prostate and pancreatic cancer. The abnormal gene may also pass down his family line. His soon-to-be-wife is pregnant with their first child. Some of Roisin's aunts and uncles also carry the cancer variant.

'I've made us sound like a very sad family,' Roisin remarked as we said goodbye, 'but we're actually very happy.'

And that is exactly how she came across. Happy, despite it all.

⋆

The idea that breast cancer could run through families was first described in 1866 by anatomist Paul Broca. He was a physician best known for locating the expressive language function in the left frontal lobe of the brain, but he was also involved in cancer research. His wife developed breast cancer at a young age and he traced the disease back through four generations of her family. He later wrote a report suggesting the disease was heritable, which was met with scepticism at the time – perhaps because the vast majority of women with breast cancer do not have a family history of the disease.

Risk-reducing mastectomy and salpingo-oophorectomy began in the early 1970s, although were very rarely taken up. Before genetic testing, a person couldn't possibly know if they'd inherited their parent's tendency to cancer, so surgery was a drastic gamble. What's more, those early mastectomies often did not remove enough breast tissue to adequately protect a woman against cancer. The discovery of BRCA genes in the

1990s, coinciding with gradually improving surgical techniques, opened the door to preventative surgery.

Risk-reducing surgery is a significant undertaking but if they could turn back time, the majority of women like Roisin would have the surgery again. That is the case even though adverse complications after surgery are common – 50% of women suffer negative effects on body image and sexuality[5] and up to 56% of women need further surgeries.[6] Mastectomy impacts psychological wellbeing and sexual relationships. Oophorectomy is associated with sexual dysfunction and mood disorders that are only partially mitigated by hormone replacement therapy. The reduction in oestrogen can also contribute to cardiovascular disease, hyperlipidaemia, hypertension, osteoporosis and depression.[7]

The negative consequences of surgery are particularly sad when you consider that it is inevitable that some women who have this surgery would never have developed cancer. No BRCA variant is associated with 100% risk of cancer and so, since no one can know who will and who won't be the lucky escapee, surgery is done based on statistics that accept that a small proportion of women will have surgery that is unnecessary in order to save a larger number from cancer. If one only considers some of the highest risk variants, like Roisin's, then it means that 10–15% of women who have had prophylactic mastectomies and 40% who have had oophorectomies would not have developed cancer if left alone. Surgery is not routinely recommended for variants with more moderate risks because the rate of unnecessary surgery would be considerably higher – more than 50% in some. Still, some women with those odds opt for it because, like Roisin, they have seen their mothers, grandmothers, sisters, aunts, cousins die of cancer. They are reluctant to wait around to find out if they are one of the ones destined for cancer or not.

Most innovations in treatment in medicine involve trials in small groups of people that lead to bigger trials and, ultimately, to large, double-blind, placebo-controlled trials. But risk-reducing surgery could never be introduced that way. There is no way to run a randomised trial comparing risk-reducing surgery with no surgery in women with a high-risk BRCA variant. That would require compelling some women to have surgery and denying it to others. Since a trial like that is a practical and ethical impossibility, surgery had to simply begin based on the assumption that it was the right thing to do. More recently women are given statistics to guide their choice, but these statistics didn't exist until recently so one really has to feel for the bravery of the women who underwent risk-reducing surgery in the early days after predictive testing became available. Back then, women who had surgery were required to take a giant leap of faith.

Thirty years of experience has allowed geneticists to improve the accuracy of their risk assessments and surgeons to perfect the various surgical approaches. But even thirty years isn't a very long time in cancer and surgery follow-up. Any innovation that is introduced without randomised clinical trials will inevitably be plagued by unknowns for far longer than that. The science of predictive medicine is very new and is immersed in uncertainty. Speaking to Roisin, it was obvious her surgery had been traumatic, but it all felt worth it to her because she was sure it had saved her life. But did it save her life? It is actually much harder than one might think to determine that preventative surgery for cancer does prolong life and that is the key point. If one operates on lots of people so that cancer never has a chance to start, it will mean that almost none of those people will get cancer. But a question always hovers over those people's survival – how many would have survived anyway without the treatment?

By definition, every person who would have survived without treatment was overdiagnosed.

The controversy around cancer screening programmes illustrates the problem.

★

Any person of a certain age who lives in a country with a good health service will be called forward for cancer screening on a regular basis. Screening is a way of identifying apparently healthy people who may have an increased risk of a particular condition. Cancer screening aims to detect cancer when it is still presymptomatic, and it mostly begins in middle age. Breast cancer screening was introduced in stages around the world from the mid-1970s onwards. So we have had many more years to understand its strengths and weaknesses than we have had for risk-reducing surgery.

Like preventative surgery for people with a BRCA variant, screening programmes are introduced without any of the stiff regulatory requirements that the development of a new drug would face. It is generally assumed that they are good for us, that they will reduce deaths from the cancer being screened and will save lives overall. On the surface, that makes a lot of sense, but here's the rub. For many different cancer screening programmes, these assumptions have been proved wrong many times over. Cancer screening programmes do not necessarily reduce *either* cancer deaths *or* overall mortality. Some only diagnose more people with cancer and treat more people for cancer without having any impact on deaths.

The first thing to understand is that not all cancer cells grow to cause either illness or death. There is a distinction to be made between cancer found on screening and cancer found because it is causing symptoms or through self-examination. The latter is cancer that has already shown signs of growth, but a screened cancer may be no more than a tiny collection of abnormal cells that can only be detected by medical investigations and which may never grow to cause symptoms. Lots of people live out their lives with exceptionally slow-growing or indolent cancers without ever knowing they existed.

The second thing to understand is that scientists do not yet know how to distinguish a slow-growing cancer from one that will cause health problems. We have only recently had technology sensitive enough to detect cancer cells at an exceptionally early stage so their natural history has never been followed up in the long term. Screening programmes just assume all cancers will grow malignantly and threaten life, so, when found, all cancers are treated equally aggressively.

A Detroit-based study illustrates this point. Autopsies carried out on men who died for a reason other than prostate cancer revealed early prostate cancer in 45% of men in their fifties and nearly 70% in their sixties.[8] The lifetime risk of prostate cancer in the US is 13%, which means most of these cancers found at autopsy were a chance discovery and would never have caused these men any significant health problems in life. But, had these cancers been found on screening, nobody could have said which were benign and which might grow malignantly, so these men would all have been subject to lots of unnecessary, invasive treatment. Prostate surgery is a major undertaking. Quite apart from anything else, it causes erectile dysfunction in one in three men.

The difficulty that every cancer screening programme faces is to find a balance between missing as few fast-growing cancers as possible (underdiagnosis) and avoiding detecting too many early cancer cells that were never destined to progress to cause health problems if left alone (overdiagnosis). That most programmes err significantly on the side of overdiagnosis is demonstrated by the outcomes of numerous screening programmes. A successful screening programme should prevent late-stage cancers, cancer deaths and overall mortality, but time and again they do not work out that way.

After the introduction of neck ultrasound in the 1980s as a screening tool for thyroid cancer, the rates of thyroid cancer diagnosis skyrocketed worldwide. But, when the rates of late-stage thyroid cancers and death rates from thyroid cancer were scrutinised, there was no change. In the US, nearly four times as many people were being treated for thyroid cancer but no lives were saved or even prolonged.[9] [10] This outcome strongly suggests that the majority of cancers found on screening did not need to be treated. This is an example of overdiagnosis by overdetection.

Breast, prostate and melanoma cancer screening programmes have all fallen foul of this problem – more people treated for early cancer but no reduction in late-stage cancer or death. A recent trial showed that an estimated one in six prostate cancers found by screening were overdiagnosed.[11] Melanoma overdiagnosis in Australia is now said to occur at 'alarming' rates of up to 76%.[12] [13] The most realistic estimates of overdiagnosis of breast cancer lie between 10–30%. Although one 2023 US-based study even suggested that overdiagnosis of breast cancer in women over seventy could be more than 30% and possibly as high as 54% in the over-eighties.[14] What a 30% overdiagnosis rate would mean in practical terms, as estimated by a Cochrane review, is that for every 2,000 women screened for cancer, one life will be saved – and ten women will have cancer treatment they may not have needed.[15] That means unnecessary mastectomies, radiotherapy and chemotherapy.

Of course, not everybody agrees that overdiagnosis of cancer on screening occurs at such high rates. If they did, screening programmes would not be so popular among health services. A 2022 study of NHS patients in England suggested that as few as 3 per 1,000 cancers detected by screening were overdiagnosed.[16] It is incredibly hard to measure the success of cancer screening programmes, which is why overdiagnosis estimates differ. The only certain way to get an accurate measure of overdiagnosis would be to avoid treating all early screened cancers and monitor them instead. But that is not easily feasible, so the issue of overdiagnosis caused by screening is a difficult one to resolve.

In general, two main outcomes are used in assessing the success of a cancer screening programme – cancer-specific mortality and all-cause mortality. Cancer-specific mortality only considers deaths from the cancer being screened but all-cause mortality considers deaths for any reason. The apparent value, or not, of cancer screening programmes depends very much on which of these two measures is used.

The majority of follow-up studies of screening programmes only look at cancer-specific mortality. That is a problem because this measure can produce overly optimistic results if not closely interrogated. An oversensitive screening programme will pick up lots of cancers that were never destined to progress. That means people will be given unnecessary cancer treatment. Treating lots of people for cancer who didn't actually need treatment artificially inflates survival rates, making it seem that people with cancer are living longer than ever before. Cancer-specific mortality measures have no way of detecting those people overdiagnosed with cancer who would have survived anyway. What's more, all those people overdiagnosed with cancer cannot possibly know that was the case so when their cancer is successfully eradicated they are likely to feel they have been very lucky. Cancer screening programmes that overdiagnose therefore come with high patient satisfaction levels and misleadingly high cancer survival rates, which only perpetuates the myth of the original assumption that early diagnosis is always for the best. The cancer screening is deemed a huge success and the programme is extended. The consequences of the chemotherapy, radiotherapy, surgery, psychological and financial burdens of a cancer diagnosis are not often captured by a measure of cancer-specific mortality.

When all-cause mortality, deaths for any reason, is applied to cancer screening, the results are usually altogether less optimistic and even a little depressing. In 2023, a meta-analysis of cancer screening programmes was published in the *Journal of American Medical Association*. Rather than looking at cancer-specific deaths, it asked if *overall* mortality improved with screening. The study included over 2 million people who had been screened for a range of cancers including prostate, breast and bowel. In the case of large bowel cancer, screening extended total lifespan by one hundred and ten days, but for all other cancers there was no evidence that finding early cancers on screening allowed people to live longer.[17]

To put this more simply, in this group of 2 million, some people found to have cancer on screening will have lived longer through receiving early cancer treatment. So their lives *were* saved. However, the *overall* survival rates (all-cause mortality) for the whole group did not improve because others were given treatment they didn't need. Some of those 'others' would have lived anyway if the cancer was never detected so their survival does not impact all-cause mortality. But some of them may have had their lives foreshortened by unnecessary treatment, offsetting those lives that were saved. For every life improved by screening just as many, if not more, may be made worse. For every life saved, another may be lost. What the all-cause mortality statistics suggest is that we do not have that balance between underdiagnosis and overdiagnosis right.

Of course, proponents of more screening make a different argument, saying that all-cause mortality trials simply do not include enough people and that is why they do not show improved survival. But that asks us to put trust in trials that have not been done and to ignore the existing evidence that many screening programmes have an absence of effect on overall mortality. The medical community is aware of the problem so screening programmes are constantly being revised and improved to avoid too much overdiagnosis, but even the most finely tuned programme has to treat some people who never needed treatment in order to save the lives of others.

Women found to have a high-risk BRCA variant who are considering mastectomy face exactly the same question about whether or not having surgery will really save their lives. Mastectomy reduces their risk of breast cancer but so far there is little evidence it actually prolongs life in the way that many women suppose it does.

Certainly risk-reducing salpingo-oophorectomy to prevent ovarian cancer has been shown to improve all-cause mortality so can be considered life-saving. Ovarian cancer spreads quickly. Screening often detects late-stage cancers that have already spread, meaning there is a low expectation for curing a cancer found on screening. So surgery is strongly recommended, with the main question being the timing. In postmenopausal women, the decision to have surgery is not so complicated because the ovaries are no longer functioning. The psychological trauma and physical complications of surgery are, for most people, offset by the benefits of surgery. Pre-menopausal women face a much more difficult decision. They must balance the high chance of developing cancer against the experience of surgical menopause and accepting that, going forward, they cannot bear their own children.

How long to delay surgery to allow time to have children is a very personal decision that each woman must make for herself. The medical recommendation is that a woman with a BRCA1 variant have a salpingo-oophorectomy aged thirty-five and a woman with a BRCA2 variant, which is lower risk, have the surgery when she is forty-five.

But the decision to have a risk-reducing mastectomy is in many ways a more difficult one. Breast cancer screening is better at finding early, localised cancers, which means that, unlike ovarian cancer, cancers found on screening are usually very treatable. This is why the decision to have a risk-reducing mastectomy is not necessarily a life-saving one.

Unlike ovarian cancer, a woman at high risk of breast cancer has options. Instead of surgery she could choose surveillance – which means intensive screening using imaging techniques and regular physical examination to detect cancer at its earliest stage. Starting around age twenty-five, or from when she tests positive for the variant gene, she will have yearly breast MRI with mammography added at between age forty and fifty. (Standard breast cancer screening involves only two-to-three yearly mammography starting after the age of fifty).

A woman deciding how to respond to a high-risk BRCA variant is more realistically choosing between the burden of cancer surveillance and living with the fear of cancer versus an immediate definitive solution in the form of major surgery. She is choosing between waiting for cancer, yearly scans and a 64% likelihood of needing chemotherapy if a cancer is found, versus having her healthy breasts removed to prevent all that. It's a question of which type of stress she thinks she can tolerate. Opting for surveillance could mean cancer is never detected, so she never has to have that mastectomy. But the stress of a lifetime of screening should not be underestimated. It could impact a woman in the same way that the fear of carrying the HD gene was all-consuming for Valentina. Added to that, one in ten MRI scans show borderline changes that may turn out to be nothing but will lead to further tests and inevitably cause anxiety. A woman starting surveillance in her twenties or thirties can expect at least a couple of false alarms from her yearly scans.

This is the most striking part of the cancer gene story for me – how new this science is and how little reliable information women have had to go on when making these enormous decisions. I have also been struck by how aggressively this problem has been addressed while we have waited for science to provide more definite answers. A woman's risk of cancer is made using statistical models. But the most accurate statistical models of risk have only been available since after 2017. Fifteen years ago, most women with BRCA1 variants were automatically advised they had a breast cancer risk of 85%, some of the worst odds. Since then, more testing in a broader range of people has led to a big improvement in prediction models, allowing more precise individualised odds. The steady rise in risk-reducing surgery began in the late 1990s, long before these individual risk profiles were devised. Many of those women given 85% odds of cancer fifteen years ago would be given a lower risk by current models. Which means more probably had surgery than needed it or would opt for it today.

There is also little good data to compare survival rates for risk-reducing mastectomy versus surveillance, and, until 2024, no direct comparison studies looking at how satisfied patients felt with the choice they'd had to make. So many women have probably been forced to choose between these two options without a good sense of whether women who had surgery or surveillance fared better.

There is evidence to suggest some women are making the decision to have surgery out of fear and there may be a social trend at play. Since 2013, when actress Angelina Jolie revealed she carried a BRCA1 variant and had undergone risk-reducing surgery, there was an increase in the uptake of genetic tests and surgery worldwide.[18] [19] Although US national guidelines do not recommend mastectomy for women with a moderate risk of cancer (two-to-four times average risk), the rates of surgery for those women has been shown to be similar to those who are at high risk, suggesting more surgery is happening than is recommended or strictly necessary. Some have raised concerns that the rise in risk-reducing surgery is unnecessary and more fear-driven than fact-driven.[20] [21]

The USA and UK have the highest rates of uptake of surgery, at roughly 50% and 40% respectively. In contrast, one German study showed only 11% of women faced with this choice opted for surgery,[22] while the uptake in France and Poland has been reported to be as low as 5%.[23] Medical practice varies between countries and cultures in many fields, but this is a striking difference and raises some important questions. Are the UK and USA performing too many surgeries or are France, Germany and Poland doing too few? The answer should be obvious in the data. If France, Germany and Poland are less aggressive in the way they manage hereditary cancers, does it mean more women are getting late-stage cancers and dying of cancer? Or, alternatively, is surveillance in those countries better so women don't need preventative surgery? How do the uncertainties of surveillance impact those women? I cannot answer those questions either way because the research hasn't been done. But if I were facing this decision, it is something I would definitely want to know. This is an area of medicine in which intervention tends to precede evidence.

There is also an interesting comparison to be made between predictive diagnoses for breast and gynaecological cancers and other hereditary cancers. Lynch syndrome causes hereditary colorectal cancer. Like with the BRCA1 variant and breast cancer, the risk of developing colon cancer is as high as 85% with some variants. But nobody seems to be suggesting a prophylactic colectomy for those at high risk and there does not seem to be a public thirst for it. The operation to prevent colon cancer is no more complicated than the multiple surgeries Roisin has undergone. Screening for breast cancer is less invasive and unpleasant than screening for colon cancer. But, in this instance, it seems, doctors and patients have decided that surveillance is preferable to surgery. Why?

All of these uncertainties are doubly important because, very soon, a great many more of us may be facing the decision that Roisin faced. A trend in medicine is that once a new technical capability becomes available, it is rolled out first to a small group and then to gradually widening populations. The assumption is that something which is of proven value to one will be of equal value to all. But our enthusiasm for new types of diagnosis doesn't always fully acknowledge that there is a learning curve as we figure out how to manage every new technique.

As things stand, the CDC and NICE guidelines for BRCA testing in healthy people are broadly similar. Currently, each restricts testing to those with a strong history of high-risk cancers in close family members. However, that's changing. Soon, people with no family history of cancer could be tested for cancer genes. It's already happening in some communities. Crucial in what follows is to understand that the risk prediction models for cancer used by geneticists weren't generated solely from looking at which BRCA variant was associated with which cancer. Also feeding into the algorithm is the person's family history of cancer. Assessing risk in people *without* a family history of cancer will likely produce a whole different set of predictions. Extending genetic testing to new groups will see us dipping our toes into some very unclear waters.

It has long been known that certain communities are more prone to hereditary cancer than others. One in forty people in the Ashkenazi Jewish population have a pathological BRCA variant. For that reason, there has always been a lower threshold for BRCA testing in a person of Jewish heritage. People from this community may be offered testing even if they do not meet all the testing criteria set out by CDC and NICE guidelines. More recently, with this risk in mind, NHS England has started to offer free genetic testing for BRCA variants to anybody over eighteen who has one or more Jewish grandparents – even if there is no family history of cancer.

On the surface, this might seem straightforward. Surely, if I had the same BRCA variant as Roisin, I would also have an 87% chance of getting breast cancer and a 60% chance of getting ovarian cancer? But the truth is not so simple. The tests for Lyme disease and the assessments for autism had to be taken in context and, as it turns out, despite the high-tech sheen of it, genetic diagnosis is just the same. All test results require a degree of clinical interpretation, genetics tests included.

Professor Anneke Lucassen is professor of genomic medicine in the NHS. I asked her about the wisdom of extending BRCA screening to women without a family history of cancer and she summed up the problem quite pithily.

'Answer me this,' she said, smiling. 'Would you recommend a whole-body MRI scan as a screening test for one of your patients?'

No, I wouldn't. Turning a diagnostic test into a screening tool is rarely wise. Findings on an MRI must be interpreted in the light of a patient's symptoms. If a patient came to me complaining of an explosive onset headache and I arranged an MRI scan of their brain that showed an aneurysm (a weakened bulge in a blood vessel), I might reasonably assume that the aneurysm was the cause. Aneurysms do cause explosive headaches, meaning the scan results and symptoms corresponded. That scan could be considered diagnostic and I could refer that person for radiological or surgical treatment of the aneurysm.

On the other hand, if a brain aneurysm was found on a scan done as part of a health check I would react completely differently to it. Aneurysms don't necessarily need to be treated unless they are very large or there are symptoms that indicate they have leaked. I would consider an asymptomatic aneurysm found on a scan done as a health check as an incidental finding. Aneurysms are a fairly common incidental finding on MRI scans and are thought to be seen in as many as 3% of the healthy population. Procedures used to treat aneurysms are very high risk, which is why treatment isn't offered automatically for aneurysms that have never caused any problems. We are all as different on the inside as we are the outside. Many of us have lumps and bumps and cysts in our organs that will never cause us problems and do not need to be treated.

In just the same way, a full clinical setting is needed for a geneticist to make sense of the result of a genetic test. For geneticists, family history is central to context. Lots of studies let us know that certain variants convey a high risk of cancer to people with a family history of cancer, but there is still little comprehensive information telling us about the impact of the same variants in people without a family history of cancer. Perhaps BRCA variants affect different people differently. In fact, there is some evidence to suggest that cancer associated variants are not equally potent in everyone.

'If you consider a population cohort like the UK Biobank, you will understand the problem,' Professor Lucassen explained. 'More people in their sixties and seventies have a high risk BRCA variant than have developed cancer. That means the 60–85% chance of developing cancer found in those with a family history does not necessarily apply to the general population. That's when you realise the genetic code isn't as predictive as we expect it to be.'

The UK Biobank is a database containing the genetic and health information from half a million UK participants. It is the most comprehensive and widely used biomedical dataset in the world and is used in a great deal of global health research. It's not a perfect set of data because most of the participants are white and are healthy and wealthy enough to volunteer for research. However, it is still a very valuable database as it has sequenced more genomes than any other study and has access to the health records of all the participants. Among that Biobank group, as Professor Lucassen pointed out, many healthy people have been found to have variants known to predispose them to specific diseases but, for some reason, these people did not develop those diseases. This finding confirmed a suspicion long held by geneticists that more is needed to develop the disease than just the disease gene.

A lot is known about disease genes, but a lot less is known about those factors that stop some people getting cancer. The human genome is the complete set of human DNA. So far, only a small percentage of the human genome has been decoded. For every gene variant that causes disease, there could be another that protects against it. The same variant does not convey the same risks in all people because there are other genetic and non-genetic factors at play, not all of which are understood.

Familial diabetes also illustrates this point. There is a gene variant that puts people at high risk of diabetes. When the variant is found in a person with a strong family history of diabetes, that person has a 75% risk of developing diabetes. However, the exact same gene variant is also found in 10% of the general population and when found in people *without a family history* of diabetes there is no increased risk.[24]

If the same proves true for the cancer gene as for the diabetes gene, it may mean that a roll-out of BRCA testing to people without a family history of cancer could produce results that are very hard to interpret. Nobody knows how many people in the general population are likely to have a high-risk BRCA variant because that level of population testing has never been done. It could be that a great many more people without cancer and without a family history of cancer carry a BRCA variant than we realise. Testing of women without a family history of cancer could mean that many more women will be offered radical surgery that they didn't necessarily need. Pre-2017, before risk models were as accurate as they are today, most women with a BRCA variant were automatically told they had an 85% risk of cancer even though later models would have given them a lower risk. Something like that could very easily be about to happen again. And it could be a self-perpetuating phenomenon. When all those women who have unnecessary surgery do not get cancer, it could be presented as a success for the BRCA screening programme – even though many of the women may never have been destined to get cancer in the first place. That might lead to a domino effect, promoting more screening programmes and more surgery. And it would not be possible for any of the women to know that they might have been overtreated.

*

'When I got the email that showed I had the BRCA gene, it was like somebody pulled the rug out from under me,' Judith told me. 'It was the worst day of my life. I tested because I really thought I didn't have it. I can be kind of paranoid medically and I thought the test would give me reassurance.'

For the current price of £129, $148 or €148, a BRCA test can be bought over the counter. Direct to consumer testing (DCT), as it is known, offers ancestry and medical tests to adults through private companies like 23andMe, Randox Health and others. These tests are sometimes marketed as potential Christmas and birthday presents. Fun for all the family. Advertisements use cheery cartoon characters to explain the science. And it's popular. By 2023, it was estimated that more than 26 million people would have had genetic tests carried out in the commercial sector.[25] There is something faintly ludicrous about my concern that genetic testing is being rolled out too quickly within the health sector when genetic tests for medical conditions are already available in the commercial sector, with none of the requirements of formal in-person consent or counselling.

Judith, a Los Angeles-based journalist, decided to have an ancestry and medical screen carried out with a commercial company on the recommendation of her physiotherapist. He thought that genetic testing might help guide her future exercise programme. Judith ticked the box for BRCA testing without much thought at all. She had a paternal aunt who died of breast cancer, but Judith was not aware of BRCA variants in the family. She received her results in an email.

'I opened the email without worrying,' she told me. 'And I thought, *Oh look! All my ancestors are from the same place.* And, *Oh look! – I like cilantro!* As I read on, I gave all the results the same weight, clicking through, not thinking very much. And I got to the end and it said – you have the BRCA mutation. BRCA1.'

DCT health tests do not describe what they do as diagnosis, but rather an assessment of genetic health risks. This allows the commercial sector to dispense with the counselling sessions and in-person delivery of results usually required of genetic testing within health services. Online educational information is provided in written and video form. There are reminders that lifestyle factors are as important as genetic factors for health. Most companies include advisories that genetic results have wider implications for family members. Some suggest a person seek out a genetic counsellor before they take a health check test. But, of course, if seeing a genetic counsellor was as cheap or available as direct to consumer testing, people wouldn't need the low-price version.

When a person sends their saliva sample for analysis, it is with the assumption that they have read, listened to and understood all the information provided.

'I just went click, click, click,' Judith admitted.

'Some websites advise people to seek counselling first. Do you recall seeing that?' I asked.

'They might have said, "Please see your doctor", but I honestly didn't pay any attention,' she said. 'I remember there were cute little characters talking about genetics. And it all had a new-tech sheen to it. I don't remember there being any sense of weight around the medical testing aspect of it.'

Judith is educated, well informed and, as a journalist, is generally inclined to do her research. But in this instance, she did what most of us do when asked to agree to terms and conditions – scrolled and ticked the required boxes without paying attention to the small print. When the positive BRCA result came back, she was unprepared for it. She is not alone. One study showed that 38% of people who submitted samples for health screening as part of DCT didn't even consider the possibility of a negative health result.[26]

With no sense of who she should contact for advice, Judith rang her gynaecologist.

'My gynaecologist said, "This is great news!"' Judith laughed. 'I mean, this was literally the worst news I have ever had but that's how she reacted. It took me a really long time to understand why she said that. As a doctor, she must have seen lots of patients get cancer and have chemo and die. I was a patient saying to her – here's my results so that doesn't have to happen to me.'

The sequencing techniques used in the commercial sector are not clinical grade and have been shown to yield up to a staggering 96% false positive rate.[27] Since this kind of testing is not referred to as 'diagnostic', a positive test for a significant BRCA variant is not supposed to be acted on as if it were. A positive result should result in a medical consultation that involves more accurate genetic testing. But one would have to read the small print to know that.

Fortunately, Judith was well supported, both by her doctors and her insurers. She met with a geneticist who counselled her and repeated the test. This clinical grade test was also positive for a pathological BRCA1 variant. Judith is of Ashkenazi Jewish heritage and her aunt was very young when she developed breast cancer, so ultimately Judith's variant was determined to indicate a genuinely high risk of cancer. Like Roisin, she felt absolutely compelled to act.

'I'm a very anxious person. I was not created to deal with monitoring,' Judith said. 'I was forty-eight, I had two children and was heading towards menopause, so it wasn't a very difficult decision to have surgery. I had my ovaries done first. That didn't feel like a huge surgery. They did it laparoscopically. It felt like my warm-up surgery. But for my boob surgery I was terrified. In the pre-op, I cried so much I was shaking. It helped for me to think that, although it was a huge deal for me, for my surgeon, it was just Tuesday.'

Learning she had a BRCA variant has also had huge implications for Judith's wider family. Her siblings, parents, nieces, nephews, aunts, uncles and cousins might all have the gene and all who were old enough needed to be advised of Judith's test result. Some have since tested and some haven't. Judith has a teenage son and daughter who will need to think about this, but not for a long time.

The right *not* to know your genetic future is easily trampled by DCT. One Canadian woman I spoke to, whose parents both underwent DCT, learned that her mother had two copies of the APOE e4 gene variant that increases the risk of dementia and that her father had one copy. That meant she would inevitably inherit at least one copy of the high-risk gene and could inherit two, giving her a guaranteed higher risk of developing dementia than the general population. She would never have chosen to have the test but her parents unwittingly took it for her.

There are also issues of privacy and confidentiality with commercial companies. Eight million of the twenty-six million people who have had DCT are said to have consented to their anonymised data being used for other purposes, but nobody really knows what that means. What have they consented to? They have given a piece of themselves away with no sense of where it will end up. Nobody knows how the data will be used or how well it will be protected from data breaches.

But Judith is grateful she had the test and then the surgery. In an odd way, it addressed a problem that predated her decision to have a genetic screen. For a long time, Judith had feared she would die young. Her mother was diagnosed with multiple sclerosis when Judith was fourteen. A maternal aunt died from Crohn's disease and a paternal aunt from breast cancer. Judith often feared that she would meet a similar fate. Getting the opportunity to take action to prevent cancer felt like a strange gift. Judith is also aware that she is fortunate in other ways. She had the resources to deal with her unexpected result. Her insurer covered all her costs, including appointments with a psychologist. She also had the skills drawn from her journalistic career to research her situation. Judith spoke to nearly thirty women who had risk-reducing surgery before she made the decision to do the same. She had heard the good stories and the bad and knew all the possible outcomes before she went ahead. Still, she worries for other women less well-resourced and for the future of commercial testing for medical conditions.

'I worry that women who don't have healthcare or a gynaecologist might tuck away the result. There should be somebody you can call. I was alone in my office when I got my result and remember thinking, *What just happened?* But at least I had somebody to call.'

Counselling is unlikely to become available with DCT. The cost would make the test too expensive to be attractive to customers. There are no follow-up studies to show how people have reacted to positive tests. Attempts are underway to bring some regulation to the sector but as things stand, it is an area of genetic testing that skirts around the rules. DCT does not only have a potentially very high false positive rate for BRCA testing, it also has a high false negative rate. There are thousands of cancer associated variants in the BRCA genes but DCT only looks for a very limited range of those. Those often seen in Ashkenazi Jews are included but many aren't.

As Judith put it, 'I learned that the BRCA mutation is like a dictionary, and that commercial testing only tests for three pages of that dictionary, and those three pages are written about Ashkenazi Jews. I was very lucky. If I was a Black woman or somebody disadvantaged by the medical community, it might have turned out very differently.'

As in so many areas of medical research, white Europeans are much better represented by current genetics knowledge than people of other ethnicities and cultures. As much as a person in receipt of a positive commercial BRCA test shouldn't trust the result, a person in receipt of a negative test has no reason to be reassured. Which begs the question – why have the test at all?

More so even than cancer gene testing, a selling point of the health screening aspect of DCT is that it forewarns people that they have a high risk of common diseases that might be at least partly mitigated by changes in lifestyle if they act on the information. It does that by assessing polygenic risk scores (PRS).

In the last several years, large studies have used the full genetic code from huge numbers of people to learn more about the contribution of combinations of variants to common medical problems. The range of genetic variants in people with a disease are compared to people without that disease so that high-risk genetic patterns can be identified. This information is then used to calculate polygenic risk scores that estimate a person's risk of developing common medical problems like heart disease and diabetes. PRS are a much less reliable assessment of disease risk than statistical models for BRCA variants because so many genes are involved.

Proponents of using PRS to promote health herald their potential to identify people with a high susceptibility rate to a particular disease so that they can take steps to protect themselves against it. The idea is that a person who learns they have an above average risk of heart disease, for example, will change their lifestyle to minimise the other factors that compound that risk. Critics of PRS say that the scores are not anything like as predictive as they promise to be. They overestimate the genetic component of diseases while failing to take into account the much more significant contribution of environment and lifestyle, not to mention all the other genetic variables that are not yet understood.

'The genetic component is only a small part of the susceptibility risk for most diseases,' Professor Lucassen told me. 'Polygenic risk scores are very topical now, but looking at people's postcodes at birth is probably as good a predictor of whether or not a person will develop cancer or heart disease as their genetic code.'

We are not slaves to our genetic makeup. Behaviour and environment can change the way our genes are expressed, which means we have some control over our genetic susceptibility to disease. Factors like diet, exercise and sleep can turn genes on and off. The genetic code itself doesn't change, but how the body reads the code does. This is referred to as epigenetics. For example, smoking can alter the function of a gene that is important in regulating cell growth, which increases the risk of cancer. But stopping smoking again can reverse that change, so the gene goes back to behaving like it would in a non-smoker. Through epigenetics, cells with identical DNA can behave quite differently by altering gene regulation. It means that the genetic code alone, the ordering of the letters in the DNA, does not have absolute control over our futures.

Most genes are susceptibility factors. They are not a cause of disease. The polygenic contribution to disease is often nowhere nearly as significant as other non-genetic factors. The technical basis and novelty of genetic testing might mislead people into giving more weight to the score than is deserved. Numerous studies have shown that as few as 10–15% of people judged at risk of a disease by polygenic risk scores will actually develop the disease. Which can only mean that at least 85% of people are labelled high risk unnecessarily.[28] To do so may be harmless if the only outcome is to encourage a person to live a healthy lifestyle. But it could be harmful if it leads to unnecessary tests and health anxiety.

Polygenic risk scores promise to advise of the risk of serious diseases like asthma, depression, coronary heart disease, high blood pressure, panic attacks, skin cancer, fibroids, glaucoma and many others. This presumes to warn people of future health problems so they can take preventative action. But how useful is that really? Most of us know that we should wear sunscreen and shouldn't smoke and should eat a healthy diet. Is a genetic test required to reinforce that message? Are PRS needed to make us do what our doctors already advise? Better public health programmes reinforcing the importance of a healthy lifestyle might be a better way to spend limited resources, rather than genetic testing that targets individuals and produces results that are hard to interpret. Furthermore, current literature has found evidence that most people given complex genetic risk information by DCT do not actually change their behaviour.[29] [30] There is also the worry that people who have been informed that their risk of a disease is low might develop a sense of invincibility that will see them neglect a healthy lifestyle they might otherwise have followed.

At a population level, PRS that look at large groups of people generate very interesting results. PRS have value in research and may have some application in the development of health policies looking at large groups. But to try to extend PRS into a clinical setting as a diagnostic test in individuals is potentially another case of science taking something useful and eroding its usefulness by applying it in the wrong context and expecting too much of it. Gene sequencing technology is moving at a much faster speed than our ability to understand what the code means.

The genetic heritage side of DCT, the reason that many people probably buy the kits, is as problematic as the health tests. So much so that a person who sends their sample to several companies will likely get several different results. One investigation carried out by the Canadian Broadcasting Corporation found that when identical twins sent their DNA for ancestry testing to a single company, each twin got a different result.

Proponents of DCT would likely say that it is paternalistic of me to say that people may not appreciate the limitations of their own results. But these results are so complex and subject to so many provisos that interpreting them is even a challenge to medical professionals.

The potential for misunderstanding results is beautifully illustrated by ex-UK Health Secretary Matt Hancock. In 2019, when he was still in that post, Hancock revealed he had taken a direct-to-consumer genetic test.[31] It showed he had a 15% risk of developing prostate cancer. He announced his result to the press as a salutary tale. He expressed his gratitude for the result, which had given him the opportunity to discuss prostate cancer with his doctor and reminded him of the importance of screening going forward. He believed the test might have saved his life. He used his experience as a platform to reveal his vision for genetic testing to be made available to everyone on the NHS, citing their ability to predict disease and save lives.

'I would never have found this out if it hadn't been for the genomic test,' he announced to the press.

But Hancock had overstated and misunderstood his result. The average lifetime risk of prostate cancer for men of his age at that time was only 12% so his personal lifetime risk was only fractionally higher than average. What's more, his risk of developing prostate cancer in the following ten years was close to zero.[32]

New technology has always captivated scientists and the public. Leaps forward suggest modernity and modern medicine tends to be perceived as superior to what went before. But new technology always comes with a learning curve. In 1895, somewhat by accident, Wilhelm Conrad Roentgen discovered the X-ray. The following year saw an X-ray craze take over the modern world. X-rays became a fun novelty. The parts needed to create X-rays were easy to obtain so people could make images in their homes, and people X-rayed anything small enough to fit in the apparatus. X-rays were seen as a potential miracle cure that could do far more than diagnose medical problems. They were credited with germicidal and beautifying properties. Doctors even employed them in hair-removal treatments. By 1897, however, people started to notice burns on the skin related to X-rays. By 1910, doctors who had developed X-ray technology were dying of cancer. There are teething problems with every new technology as we figure out how to use it. That is the case even if that technology is revolutionary and has lasting value to medicine. X-rays changed healthcare and retain their value today – now that we have learned how to use them safely.

I have started to wonder if we are in our own '1896' for genetic testing. Its use has exploded both inside and outside healthcare services. The ability to do cheap and easy genetic tests will certainly revolutionise medicine, and the value of genetic diagnosis and screening will be sustained into the future. The more genetic tests that are done the more reliable the results will become and the more useful it will be. But we're not there yet, even if our behaviour suggests we think we are. There is a much bigger gap between our ability to sequence the whole genome and our ability to interpret the results than people realise.

I also worry that we are so pleased with our ability to detect biological abnormalities and so impressed by technology that we have become like kids in a candy store. Some things in medicine are relatively easy to do and some are very difficult. Finding cancer is considerably easier than understanding why it progresses differently in different people. Finding genetic diseases is considerably easier than treating them. Finding genetic variants is much easier than understanding what they mean. It is very tempting to do the thing we have already mastered over and over as we wait for the slow thing to come to fruition. But that doesn't allow for the fact more tests and more diagnoses do not necessarily make for healthier people and that diagnosis has a power all of its own that can make us sick.

More people are dying from cancer and more young people are being diagnosed with cancer than ever before. These are true cancers, not found on screening, that have come to light through symptoms. So cancer should be a big concern for society. It is certainly a challenge for scientists and doctors. While treatment for cancer is much better than it was, there are still a lot of lives that need to be saved. Globally, nearly 700,000 women die of breast cancer every year.[33] But we need to do a better job of distinguishing those cancer cells that will progress from those that won't. We need a better understanding of which factors make cells grow into cancer and which protective factors stop that from happening. Until we manage that, an undetermined number of people will be exposed to toxic treatment that they don't need.

Fear is a powerful driver of overdiagnosis. Cancer is something that scares people and compels us to act, professionals and the public alike. But sometimes it is OK to stop and think for a while. There are watchful waiting programmes for some early cancers found on screening.[34 35] This means regular check-ups and time to think. But to allow ourselves that luxury, there needs to be much more public clarity and openness about the cancer overdiagnosis challenge associated with screening programmes. We also need a doctor we can rely on to guide us through a waiting game like that. Facing the prospect of cancer is not something anyone should do alone or in an atmosphere in which they do not feel well cared for.

Some say we have a language problem.[36] Small errant cells found on screening that show no definite evidence of growth are given the exact same name as fast-growing, symptomatic tumours: cancer. There is an argument for giving highly localised abnormal cells a different name so that people are not frightened into action. Doctors are frightened too, to the point of leaning heavily in the overtreatment direction. In the words of health journalist Edward Davies, 'The fear of both patient and doctor can sometimes override the best knowledge, research and information known to man.'[37] The antidote to fear is knowledge, trust and support.

I do attend screening when invited and I recommend others do so, too. If someday I am the unlucky one to have a positive test, I plan to allow myself space to consider my options. But I also remind myself that screening is the least of the things I can do to protect myself from cancer. Smoking, obesity, alcohol, diet and sun exposure are all factors that increase the likelihood of getting cancer more than genetic risk. The things I do to limit these risk factors are what will give me the best chance of living a longer and healthier life.

5
ADHD, Depression and Neurodiversity

School was a difficult time for Anna. It still haunts her. She recalls being a sociable child, good at making friends. But she also remembers becoming hyper-fixated on one friend, then another and another in succession. She was accused of using people. She tended to be impulsive and, wanting to please others, was easily led. One distressing incident in particular has never left her. She had moved school in the middle of an academic year. On the first day at her new school, she was relieved to be taken under the wings of two girls. At lunchtime, in fits of giggles, the girls egged each other on to do naughty things. Anna spat orange juice at the boys. She did it with relish only to reproach herself later. She feels the episode coloured her whole school experience.

'What must everybody have thought of me, doing that on my first day?' she told me tearfully.

As a child and as an adult, Anna felt sanctioned, judged and misunderstood. She considers herself a chameleon who adapts to new environments and survives by being funny, but all too often regrets things she has said. Her self-esteem is low. After school, she went to art college, but somewhere along the way decided she wasn't good enough to be an artist. She needed to see success in order to keep going and, despite receiving commissions and having paintings featured in exhibitions, she couldn't. So she became a nurse. Her mother was a nurse so she had a model for what that looked like. The clear rules and guidelines that came with the job appealed to her.

'Actually, it was the perfect job for me,' she told me. 'Every day is different. I thrive off being with other people. I love making people feel better.'

Although Anna loves her job and is good at it, she still often feels inadequate.

'People don't think I'm as clever as I feel. I can't get the words out quickly enough,' she said.

When she was in her early twenties, she remembers being exhausted all the time. She was overwhelmed by work. She made mistakes that she couldn't admit to others.

'I worked until late at night to cover my tracks,' she told me.

She was tired but couldn't sleep. She couldn't control her emotions. She complained bitterly of a poor memory. She left her keys in ridiculous places, one time in the fridge. She forgot to turn off the hob and to unplug the iron.

'I can never remember things people tell me, like how many children they have. I hate it because it makes me look like I don't care and I do.'

Anna has sought medical advice many times throughout her life. In her twenties, her GP diagnosed her with depression and put her on antidepressants. They made her feel numb so she stopped them. She saw a nutritionist who told her she had a yeast infection and put her on a restrictive diet. It helped but only for a few months. She saw a therapist weekly for ten years and found that beneficial. Still her mood fluctuated. She had bouts of severe depression broken up by happier times, during one of which she met her now husband, Malachy.

Anna only came to wonder if she was neurodivergent when she was in her forties, after a conversation with a friend who had been diagnosed with ADHD.

'I have always been lovingly teased as somebody who would trip over the pattern in the carpet,' she smiled. 'I was quirky. So when my friend sent me a newspaper article about neurodivergence, the description hit me like a bus. The article could have been written about me.'

After some research, Anna paid to have an online assessment for ADHD. This was during Covid lockdown when face-to-face assessments were not available. The diagnostic process required Anna, Malachy and Anna's mother to fill out questionaries in advance of the consultation. The diagnostic interview took ninety minutes. Anna found the assessment very thorough. The assessor, who Anna thought was a psychiatrist but wasn't sure, drew her attention to things she hadn't noticed about herself. When asked if she was excessively fidgety, Anna replied that she was not, but the assessor pointed out that she had twiddled with her hair repeatedly during their consultation. Anna realised that she *was* unusually fidgety, remembering how she always doodled and played with her pen during meetings.

'He asked me if I'm compelled to get up and leave mid-meeting. I said no to that too, but when I thought about it more I realised I wouldn't allow myself to get up and walk out. I suppress the impulse so much that it doesn't exist. I mask.'

Anna now realises she has underplayed her ADHD traits all her life. The diagnosis made so much sense to her. Slow neurological processing explained why she had such a bad memory and could not communicate her ideas to other people as clearly as she wished to do. No wonder she was exhausted with all the effort required to camouflage her difficulties and hide her mistakes.

Anna now takes the stimulant drug methylphenidate (Ritalin).

'The first time I took it, I couldn't believe how clearly I could think.'

After a few adjustments to the dose, Anna noticed a distinct improvement. She could make decisions more quickly and prioritise. She had more energy.

I wondered how this impacted the practical aspects of her life, in work and in her relationships.

'Is my life better? Yes and no,' she said.

Anna's workplace has been supportive. They have made adjustments, such as giving her a private, low-ceilinged office which she feels reduces noise reverberation and provides quiet when she needs it. She is also allowed to wear noise cancelling headphones if she is in a busy environment. Colleagues have been taught not to burst into her office without warning. Interruptions make her angry if she is focused on work. She has a disability passport that helps communicate her difficulties and special needs to others. Despite this, Anna still finds work challenging and feels she has to constantly remind people that she has a disability. So much so that she is off work now and does not foresee herself ever returning. Part of the ongoing problem is that neither Anna nor her employer know what she really needs.

'When somebody asks if I want help, I don't know what to ask for,' she told me.

Taking time off, albeit as sick leave, is actually a sign of Anna being kinder to herself. She is better able to recognise what she cannot do and to allow herself not to do it. Whereas once she would have forced herself to go to a crowded party that she knew she would hate, now she doesn't go and asks to meet friends in a setting that suits her better.

I asked how her family and friends have received her diagnosis. A common response when she describes her symptoms is 'everybody has that', I learned. Everybody is a bit disorganised. Everybody finds work hard.

'The difference is that I feel like that all day, every day,' Anna explained. 'I never feel OK.'

*

ADHD started out as a defined medical condition in the DSM-2 in 1968, in which it was referred to as hyperkinetic reaction in children, described in a single line as distractibility and restlessness that went away in adolescence. In 1980, the DSM-3 introduced the term attention deficit disorder (ADD), with hyperactivity being added to the name in the revised edition of the DSM-3 in 1987, making it ADHD. The DSM-5 describes ADHD as a pattern of inattention or hyperactivity that interferes with social functioning or development. The diagnosis requires that the difficulties be present before the age of twelve, that they are present in lots of settings, that they reduce the quality of social, academic or occupational functioning. The distinction between mild, moderate and severe ADHD is very vague. The diagnosis of severe ADHD requires 'marked impairment'. Mild ADHD comes with 'no more than minor impairments'. The DSM describes moderate ADHD as causing impairment 'between mild and severe'. There is no consensus on what counts as impairment.

ADHD is a diagnosis usually made in children. Being diagnosed with ADHD for the first time in adulthood has only become possible in the last several years. Like all medical problems, ADHD has a range of severities. One of my most vivid encounters with it was not with one of my patients, but in spending time with Kendra, who is the young daughter of a friend of mine. Aged eight, Kendra, who had severe ADHD, had a level of energy and lack of focus that was impossible to keep up with. She pinged from person to person in a group, speaking quickly and jumping from topic to topic. She was a wonderfully affectionate child and I had the sense she was also very intelligent, but it was hard to test that theory because she couldn't apply her attention to anything for long. I once joined her and her mother for a shopping trip. Such was her energy that no one could relax. Everything took her interest but never for long. I don't know how her parents kept track of her. I was sure she'd dash away and we'd lose her. She didn't. She grew up safe and well cared for to become a creative adult. She still has problems with lack of focus. She found education hard, but she found her place. She's an artist, which allows her to work at her own pace, in her own way, and which makes the most of her flexible thinking.

Diagnostic rates of severe ADHD, as Kendra had, are actually quite stable. Those with severe ADHD are now vastly outnumbered by those with mild ADHD. As is the case for autism, there have been staggering increases in the number of people diagnosed with ADHD in the last thirty years, but that growth is almost all at the milder end of the spectrum. The average global prevalence of ADHD in children is 7%.[1] In the US, ADHD diagnoses increased from 6% to 10% in children between the 1990s and 2016.[2] In the UK, they doubled in teenagers between 2000 and 2018.[3] Germany, a country with generally low rates, saw a 77% increase in ADHD diagnoses, from 2.2% to 3.8% between 2004 to 2013.[4] One meta-analysis showed a prevalence of 14% in Tunisia and 22% in Iran.[5] More than 85% of children with ADHD worldwide are in the mild or moderate group. It is so often in that grey area of diagnosis, where there is a struggle to draw the line between 'normal' and 'abnormal', that diagnostic inflation happens.

However, the biggest relative rise in diagnoses of ADHD is not in children but adults. New diagnosis in adulthood has gone from being very rare to as high as one in twenty adults in some places.[6] [7] These almost all fall in the mild domain. The UK has seen a 400% increase in adults seeking an ADHD diagnosis between 2020 and 2023.[8] Gradual adjustments to the diagnostic criteria in the DSM have made adult diagnosis increasingly possible. 'Hyperkinetic reaction of children' that disappeared in adolescence has passed through many iterations to become ADHD, diagnosable at any age, as it is now.

Many of the same uncertainties and controversies that applied to autism also apply to ADHD. There is diagnosis creep with subtler and subtler symptoms constituting a diagnosis. The diagnosis of ADHD is somewhat less formalised than for autism. It involves a detailed clinical evaluation by a qualified professional. In addition, various rating scales are available to help quantify the number of symptoms present that might indicate either inattention or hyperactivity. Many of these rely on self-reported symptoms. That makes the diagnosis inherently subjective. Also, different professionals will tend to diagnose it differently because the symptoms are so qualitative and hard to measure. In an ideal scenario the assessment should take place over a series of meetings and should involve more than one diagnostician. The DSM lists examples of the sort of difficulties that might be expected, including 'often loses things', 'often does not listen when spoken to directly', 'often avoids tasks', 'often talks excessively', 'often fidgets'. The word 'often' is open to interpretation. A diagnosis requires that symptoms interfere with the quality of social, academic or occupational functioning. That is a very difficult thing to measure. Presumably any person who seeks out an assessment for ADHD will only do so because they are struggling in some aspect of their life, which means that all will be considered impaired.

There are interesting social trends in ADHD diagnoses that speak to some of the issues with diagnosis and potential over-diagnosis. The youngest children in a class have repeatedly been shown to be more likely to get an ADHD diagnosis than older children,[9] which suggests that for some, immaturity may be being conflated with a neurodevelopmental problem. There are also substantial differences in diagnostic rates within single countries that are not easily explained by differences in culture or access to healthcare. In Norway, where there is free healthcare, readily available to all, diagnostic rates vary from below 1% to more than 8% in different regions.[10] In the US 14% of children in Mississippi are said to have ADHD, but only 5% in California.[11] This suggests that some practitioners are making substantially more diagnoses than others.

A large number of people who have ADHD also have one or more other related diagnoses, like autism, anxiety and depression. One study found that 87% of adults with ADHD had a second psychiatric diagnosis and 56% had a third.[12] The DSM-5 allowed a diagnosis of both ADHD and autism in the same person for the first time in 2013. Before that, these diagnoses were mutually exclusive. Since the DSM-5, the number of people with both diagnoses is rising steadily. Poppy, who we met in chapter three, was diagnosed with autism aged twenty, and also has ADHD, depression and an eating disorder. Anna has a history of depression and is currently considering asking for an assessment for autism.

ADHD and autism have some overlap. They both lead to social difficulties, with irritability, anger, slow processing of information, intense interests and preoccupations, a tendency to burst in on conversations and failure to work with social cues. That these overlapping diagnoses co-occur can be explained in different ways. Many believe that this is to be expected because neurodevelopmentally divergent brains are at risk of multiple different mood and behavioural problems. But perhaps more likely is that these disorders are so poorly defined that the same people are getting multiple diagnoses to explain the same difficulties.

The DSM lists ADHD as a neurodevelopmental disorder. Something interesting happens to a medical problem when it enters the DSM. Although only born out of consensus by committee rather than through scientific advancement, it is suddenly made to seem scientifically concrete. A DSM listing immediately makes a disorder feel like it is a discrete entity. That turns it into a field of study for researchers and an intensive hunt for biological mechanisms begins. Once a genetic association has been identified in some families and brain differences found in a subset of sufferers, the condition is considered ratified and its place as its own category in the DSM endorsed. The development of expert diagnostic and treatment services follow, support groups emerge, and the disorder is given longevity.

When the DSM called ADHD a 'neurodevelopmental disorder', whether it meant to or not, it created the impression that this was a singular, biologically driven, brain-based developmental disorder. That impression is reflected in public conversation. The online magazine ADDitude refers to ADHD as 'a neurological disorder that impacts the parts of the brain that help us plan, focus on, and execute tasks'.[13] The charity ADHD Ireland refers to ADHD as 'a medical/neurobiological condition in which the brain's neurotransmitter chemicals; noradrenalin and dopamine do not work properly'.[14] But what is the evidence that ADHD is actually 'a medical/neurobiological condition' or a 'neurological disorder'?

Studies have certainly shown structural brain differences, such as slightly smaller brains or brain regions, in children with ADHD compared with healthy control groups. What's important to understand is that these are not 'abnormalities', just differences seen only on comparison between groups. Radiologists cannot diagnose ADHD on a scan because people with ADHD have normal brain scans. Brain scan differences like these are often considered evidential support that ADHD is an inherent brain developmental problem but really, they shouldn't be. Most showing positive findings involve only small numbers of people, are carried out only in children and are not reproducible. While they provide interesting information that helps point to future areas worth studying, they do not prove that ADHD is a brain 'disorder' or a discrete medical condition. Neither are brain differences necessarily a 'cause' for ADHD. They are only an association. So, for example, traumatic life events and deprivation in early life can affect brain development and lead to behaviours that could be diagnosed as ADHD. Structural brain changes there would be associated with ADHD rather than causing it.

Some studies also show different patterns of brain activation, as measured by the brain's blood flow or oxygen uptake, in people with ADHD compared with healthy controls. Again, this is mostly in children and it is not an abnormality but a difference. All differences between people have some biological correlate. Brain activation studies can also detect differences in people based on thinking styles and personality traits – and, for example, between people who are night owls and early birds. What differences in brain function between a person with ADHD and a person without ADHD actually say is that the inattention or hyperactivity is real, but not that it is due to a brain disease, or that it is 'abnormal' or that it has a single cause.

Genetic studies are also used as part of the proof that ADHD is a discrete biological 'disorder'. Much of this comes from studies on twins which have shown that ADHD is 76–88% heritable. However, that has not translated as hoped to the discovery of a set of genetic variants that can potentially explain ADHD. In fact, a meta-analysis of genome-wide association studies in large populations has shown only 22% heritability.[15] The gap between this finding and that seen in twin studies means that somewhere along the line heritability is lost, but how? It's likely that the interplay between multiple genetic variants and many other non-genetic factors are required for ADHD to develop. The genetic contribution of any single variant may be tiny, with environmental factors mattering a great deal more. Neither are those gene variants that have been associated with ADHD specific to it – they are also seen in people who do not have ADHD and in other disorders. Which means the variant alone cannot cause the disorder. ADHD is likely polygenic, meaning, like for heart disease and diabetes, the genetic component may still prove to be the smallest part of the picture and a person's early life circumstances may say more about who will get it than any medical test.

Another popular theory is that ADHD is caused by low dopamine levels, but again there isn't enough evidence to say that is the actual cause. The research that supports this theory often involves low numbers of people and there are studies that say the exact opposite, showing no dopamine dysregulation in people with ADHD. Often, when a study finds some biological point of interest associated with neurodivergent conditions, that finding seeds itself into conversation even before absolute proof is established.

None of this is to say that ADHD is not biological and that further biomedical research is not needed. Of course it is. But biological does not automatically equal disease or disorder. Every mental health disorder, but *also* every feeling, every personality trait, every passing thought and every bodily change, be it considered 'normal' or abnormal', is biological. Every single ordinary human experience is reflected in some change in the brain. Neural correlates have been found to people's preference for certain chocolate brands. That ADHD has genetic determinates is not in doubt either. A person with a first-degree relative with ADHD has a five- to ten-fold higher chance of having ADHD than other people. But genetic associations and heritability do not necessarily speak to it being a 'disorder'. Genome-wide association studies can also be used to predict whether or not a person will like coriander (cilantro). Biomedical research is useful both in understanding pathology and also in determining how the brain generates human traits like personality and disposition. But it does not necessarily mark ADHD as a neurological problem, a single disorder or a primarily brain disorder as is often suggested.

ADHD is also subject to the same research problems as autism. As much milder versions of ADHD are identified, the group of people with the label becomes so heterogenous that biological commonalities are harder to find. It may be that children with severe ADHD, those with obvious problems since pre-school years that are bad enough to make it impossible for them to function normally in school, *do* have a discrete condition for which there is a *single* strong genetic association. But that will become harder to prove as the ADHD community draws in new populations.

The frank truth of it is that despite decades of work, no biomedical research project has succeeded in finding any brain abnormality common to ADHD sufferers. There are no biomarkers that allow behaviours exhibited by people with ADHD to be distinguished from other disorders or even from normal human experience. Even those researchers intent on finding the biological 'cause' for ADHD will admit it is a condition that manifests in many ways in a wide range of people and has lots of different long-term outcomes. Yet, we continue to gather people who have the traits considered consistent with ADHD under one medical category, studying and treating them as if they were a single group, all of whom unequivocally have a brain development disorder.

The biologising – or more accurately the pathologising – of mental health problems and behavioural disorders is a trend right now, both within medicine and in society. It is not unusual to hear depression described in terms of a serotonin deficiency rather than as a reaction to life circumstances. Poppy explained her inability to force herself to do essential tasks through a deficit in the neurotransmitter dopamine. It gave her a great deal of relief to conceptualise her difficulties in that way.

Finding biomarkers, such as genetic correlates or scan 'abnormalities', that prove mental health problems and psychological suffering are 'real' is popular now in part because it validates and gives substance to people's difficulties. In that atmosphere, concentrating too much on the social or psychological aspects of illness is increasingly referred to as medical gaslighting by some patients – it is seen as denying somebody else's truth. The words we use to describe our suffering make a huge difference to how it is perceived. That may be why ADHD, autism and depression are now commonly referred to as neurodivergent conditions.

The term neurodiversity was coined in 1998 by Australian sociologist Judy Singer. It is not actually a medical term but it sounds like one, so has a biologising, pathologising effect. Singer described how she came up with it in an interview: 'I got it from a combination of biodiversity which is a political term saying it's good to have diversity in the environment. I noticed that psychotherapy was becoming a bit of a joke and neuroscientists were the new priesthood, so I thought, let's put them together. Neurodiversity sounds really important and it will legitimise our claims to be taken seriously.'[16]

In my own field of work, there is also a strong trend towards looking for a biological cause of human distress. These days, psychosomatic disorders that manifest with neurological symptoms are called a functional neurological disorder (FND). Seizures that might once have been called psychogenic or psychosomatic (or, decades ago, hysteria) are called functional seizures. For many, the newer terms are regarded as preferable because they remind people that the psychosomatic disability is a function of brain processes. The use of words like 'psychological' are increasingly frowned upon in some medical circles because, even though they are not meant this way, they can be mistaken to imply that symptoms are purely psychiatric, which leads to a dismissal or underestimation of the very real physical disability. To refer to a medical problem as 'psychological' immediately downgrades its importance in lots of eyes.

As a doctor, I am a 'psychologiser', some might say. I am one of a group of doctors who prefer to ask the question, 'what has happened to you?' instead of, 'what is wrong with you?' Many psychosomatic disorders start in the context of a major life stress. The physical symptoms are the consequence of *real* biological physical changes, but they are not due to a disease and would not have occurred without the stress trigger. Sometimes in the context of life's difficulties, the physiological fight and flight pathways are so active they produce an array of bodily changes that people mistake for a primarily physical illness. When I talk to my patients about psychosomatic disorders, I usually concentrate more on the external event that set the train in motion than I do on the internal physiological process that kept it on its track. Some of my colleagues do the opposite, tending more to biologising. They might understand the patient's symptoms in the exact same way that I do, but they prefer to frame their explanation around the biological changes in the brain and body rather than giving external stressors a central role in the condition.

As a 'psychologiser', I hope that helping people to understand that something in their social environment has made them sick will give them control over how to deal with external stressors in the future. I fear that a view that talks too much about internal biological processes makes people passive victims of their medical disorder, which takes away their control. 'Biologisers' would say that placing too much emphasis on social and psychological triggers risks making patients feel they are being blamed for being sick. Neither approach is wrong. They are just different. The best approach is to tailor every conversation to the individual patient.

I do of course think that understanding the biology of all conditions is absolutely necessary, but my concern arises from how too much focus on biomedical research can distract from the many psychosocial factors invariably associated with mental health and behavioural disorders. The innate biology which creates vulnerability to mental health problems cannot always be changed, but the social and environmental factors can be. In the case of ADHD, childhood abuse, neglect, multiple foster placements, witnessing violence and alcohol exposure *in utero* have all been shown to increase a person's risk of developing ADHD. These areas of life are readily available for social and psychological intervention but they are not currently at the fore of the ADHD discussion. The DSM-5 expressly notes that signs of ADHD may be completely absent 'when the individual is receiving frequent rewards for appropriate behaviour, is under close supervision, is in a novel setting, is engaged in especially interesting activities, has consistent external stimulation'. This suggests that better lives for children will lead to fewer neurodevelopmental problems in children and, later, in adults. But this aspect of ADHD tends to be neglected in public conversation, perhaps because it is a much more delicate and difficult one to have than one about brain chemistry.

Viewing ADHD as a neurochemical disorder has also been an obvious boon for the pharmaceutical industry and that cannot be ignored. If ADHD is indeed due to a chemical imbalance, that makes it available to chemical treatment. ADHD diagnosis inflation is partly driven by the availability of stimulant 'treatments', such as methylphenidate (Ritalin) and dextroamphetamine (Adderall). There was a seven-fold increase in prescriptions for ADHD in the UK in the last ten years[17] and a ten-fold increase for ADHD in adults in New Zealand between 2006 and 2022.[18] Stimulant prescriptions increased by 250% in the US from 2006 to 2016.[19] A similar growth in prescriptions has been seen worldwide. ADHD is a business as well as a medical disorder. Stimulant drugs would never be recommended to counteract the effect of childhood adversity but they can be given if ADHD is viewed as a neurotransmitter imbalance in the brain.

Stimulants are first-line treatment for adults, while in children they are reserved until educational and behavioural support have failed. Among the working group advisers for the development of the category of ADHD in the DSM-5, 78% disclosed links to drug companies as a potential financial conflict of interest.[20] Pharmaceutical funding is also common in educational, professional and consumer websites providing information on ADHD. Drug companies also provide funding to patient advocacy groups and ADHD charities.

The irony here is that the jury is out on whether or not medicating ADHD in adults actually works. There are studies that suggest stimulants are effective, but those have mostly been short-term studies, lasting less than fourteen weeks, with many having been carried out by a single research group – so they have not yet been replicated by other study groups to further confirm the findings. Meanwhile, a 2022 Cochrane review found no good evidence that methylphenidate was superior to placebo in adults. The review incorporated 24 trials involving 5,066 people with ADHD and found that most research suggesting the drug was beneficial was of low quality.[21] Where drugs may work is to help people focus on repetitive, mundane tasks. But that does not necessarily translate to an improvement in creativity or in the sort of complex mental abilities that matter more to relationships, work and success.

*

Forgetfulness, lack of motivation, noise intolerance, social anxiety, low mood, distractibility and concentration difficulties are all part of the human experience. Each of these have become increasingly pathologised, in part because of their inclusion in categories in the DSM. It's very easy to criticise the DSM for its role in pathologising 'normal' and for encouraging the medicalisation of all human suffering, but actually I think it would be hard to do without it. Systems that classify illness and disease are essential, and I can't see how health services or research facilities would operate in their absence. For example, scientists couldn't study people with symptoms consistent with schizophrenia if those affected were so variously labelled that they couldn't be grouped together. A shared language is needed to allow professionals to communicate about patients and to develop services. Insurance companies demand diagnostic categories like these to prove we are really ill.

The problem with the DSM is not that it exists but that it is taken more literally than intended. It also seems to be very hard to dial back DSM categories, even when it is clear they have gone too far. Whenever a tightening of diagnostic criteria risks taking a diagnosis away from some people, a new diagnostic label is usually created so nobody will be without a diagnosis. That is what happened in the DSM-5, when the category of 'social (pragmatic) communication disorder' was developed to account for people who might no longer qualify as having autism according to the newest criteria. Every new volume even suggests problems worth considering as future 'conditions'. The DSM-5 committee suggested 'caffeine use disorder' as a potential category for the next edition. This is defined as problematic caffeine use leading to impairment and distress. And the biologising is already underway: studies on twins show caffeine overuse is heritable and variants in the ADORA2A gene have been associated with it.

Patterns in medicine repeat themselves, even when the original pattern hasn't gone so well. Depression also falls under the banner of neurodiversity. Before ADHD, mild depression followed a somewhat similar trajectory, becoming a chemical disorder of the brain that warranted drug treatment. When the first antidepressant, Imipramine, was developed in the 1950s, the pharmaceutical company who developed it was concerned there was no market. They shouldn't have worried – this class of drug later came to be one of the industry's big earners.

Like with ADHD, the success of chemical treatment for depression depended on attributing low mood to a chemical imbalance in the brain. In 1967, British psychiatrist Alec Coppen proposed low serotonin as a cause for depression. The idea caught on and, decades later, most antidepressants are still focused on raising serotonin levels. And yet a 2023 meta-analysis published in *Nature* concluded that the serotonin theory may have been wrong all along.[22] That analysis found no evidence for an association between low serotonin and depression. There are some studies that do show an association, but even those do not prove that the chemical imbalance causes depression. What is generally accepted now is that while low serotonin levels *might* trigger depression in *some* people, it is not the cause of depression in the majority.

But if low serotonin is not the cause for depression, then how could drugs that raise serotonin levels have been making people feel better for so many decades? Actually, there is no firm answer to that question but, for many, possibly for the majority, the improvement may be through the placebo effect. There is a growing view that antidepressants have no benefit beyond placebo in those with mild depression.[23] [24] Not that improvement through the placebo effect is negligible. Feeling better is all that matters, whatever the mechanism, and some would say we should try to harness the power of placebos in a more formal way in medicine. Although not necessarily through chemical treatment.

The use of antidepressants in severe depression is better supported by evidence, although some have even called that into doubt.[25] The science of depression, like that of autism and ADHD, has been severely muddied by the gradual homogenisation of the disorder, with severe and mild depression being treated as if they are the same, making it harder to determine what treatment works. Depression has drawn in milder forms of sadness, making it increasingly possible to refer to any low mood in medical terms. Understandable human sadness is being increasingly biologised. The DSM-5 made changes to allow a diagnosis of depression within two weeks of a bereavement. These are well-intentioned changes designed to ensure that nobody who really needs it is excluded from medical care, but they can also pathologise appropriate human reactions to difficult circumstances.

We all have periods of low mood. Depression that lasts for less than two months, which is not associated with suicidal intent, does not necessarily need medical treatment. Most cases of mild to moderate depression and grief resolve spontaneously without medical help. Medicalising them risks unnecessary medication or even slowing down recovery by focusing too much on symptoms and interfering with normal coping strategies. It also threatens the quality of research findings by treating lots of people with very variable levels of suffering under a single medical label as if they are all the same when they are not.

One in five young people in the UK is now said to have a mental health problem, which is an astonishing statistic.[26] But does the sudden growth in people with a mental health diagnosis mean we are all becoming sadder and more troubled? Or that we are better at diagnosing mental health conditions?

Or might we simply be more inclined to pathologise normal human emotion?

A recent cohort study that spanned twenty years, from 2000 to 2020, has tried to address these very questions. It looked at the electronic health records of 29 million UK residents over the age of sixteen. The results seem to lend some support to the narrative that says mental health problems are genuinely on the rise, with the biggest increases in this cohort being in 16–24-year-olds.[27] What this study certainly confirms is that more people are going to their doctors with mental health complaints. However, that still leaves the question of whether or not we are more anxious and depressed, more willing to go to our doctors with these sorts of symptoms or more likely to refer to low mood and anxiety as medical problems. Doctors may also be more willing to record symptoms as mental health problems that might previously have been referred to in a different way.

The study did find some evidence to support the view that there has been an increase in actual depression and anxiety symptoms over time – meaning people are more symptomatic rather than just referring to ordinary low mood as depression. But that also has to be interpreted with caution because we live in an era in which we are encouraged to pay attention to our feelings and in which we increasingly use medicalised language to refer to feelings. Awareness campaigns may drive people to be concerned about symptoms they would have downplayed before and may encourage them to go to their doctors with problems they would never before have considered medical. For example, the fear of public speaking is now diagnosable as a social anxiety disorder.

What is clear is that we shouldn't rush to conclusions. The jury is still out, even in matters that are often held in common public agreement – such as the assumption that one of the major reasons that young people are being diagnosed with mental health problems is down to the use of social media. The interaction between mental health and social media use is much more complex than is often presented to us, with assumptions often failing to consider the positive things social media brings to young people's lives, such as entertainment, educational information and connectedness. In fact, one study suggested social media improved self-esteem through peer support, enhancing connection.[28] Another showed that humour on social media relieved stress during the pandemic.[29] Furthermore, a recent UK-based study of 3,000 10–15-year-olds showed no link between mental health problems and social media use.[30]

Where some association has been found between social media use and depression, it could also be that people with pre-existing mental health problems are more likely to use social media, so the relationship between the two may be the opposite to what we expect.[31] It may be that certain aspects of social media use have a negative effect and others have a positive effect. Cyberbullying and a dependence on numbers of 'likes', 'comments' and 'followers' might certainly have a detrimental effect on wellbeing.[32] Excessive time spent on social media might interfere with other aspects of life. Young people who spend a lot of time on social media are less likely to play sport. Certain types of use could also affect body image and cause people to suffer through comparison. But some experts have estimated that social media use actually only accounts for 10% of the increase in mental health problems in young people and 5% can potentially be explained by reduced participation in sport. A further 5% can be explained by a decline in parents' mental health. That leaves 80% unexplained,[33] which gives plenty of room to allow for the pathologising of ordinary human experiences as a large contributing factor.

The question of whether or not there is a mental health crisis among young people is open to many different answers.

We should not assume that social media is central to the problem, but we should assume that a person must be suffering if they feel badly enough to go to a doctor for help. The question I come back to is whether or not medicalising that suffering by giving it a diagnostic label and biologising it by referring to it using genetic and brain-related explanations is the best strategy to make people better? In the past psychological suffering was neglected, so better acknowledgement and support services for struggling people should be only great news. However, as always happens when diagnostic inclusivity draws in milder cases, the negative impact of medicalisation, such as the labelling effect, is inevitably ignored. Medicalisation through biological theories and treatment with chemicals can also all too easily sideline essential social change. There is lots of evidence pointing to life circumstances as major risk factors for mental health problems. Adverse events in early life, family history of mental health problems, negative family environment, bullying, poverty, racism, social isolation and inequality are all risk factors for mental health problems. They are also possible to change so maybe that is where our attention should be focused.

The availability of medical explanations and chemical solutions for sadness have been very tempting for all concerned. A consultation that ends with a prescription is a relatively easy one for a doctor. It manages their own anxiety and makes them feel useful. It is also easier for the patient. There is something to be said for the quick fix, placebo effect of antidepressants that foreshortens a difficult conversation. But chemical treatments come with side effects, like drowsiness, agitation, insomnia and sexual dysfunction, even increased suicidality in young people on some antidepressant drugs. And while explanations for distress that rely on internal biology might offer relief, they also risk robbing people of control over their own futures and distracting from the need for social change.

The DSM approach is not the only way. In fact, many fear, as I do, that biologising mental distress and behavioural problems can get in the way of an examination of life and society that might lead to personal insights that could contribute to more lasting improvement. There is a growing feeling among some medical professionals that there should be a move away from an overly biologising approach to psychology and psychiatry. Vocal among that group is psychologist Lucy Johnstone, who sees the diagnosis of mental health conditions as obscuring personal meaning, damaging to personal identity and removing agency. Johnstone prefers to conceptualise mental health problems as survival strategies rather than brain disorders. In this theory, experiences that are described as 'symptoms' are actually a reaction to threats and a manifestation of what a person needs to do to overcome that threat. Humans are essentially social beings. Troubled behaviour and low mood are inseparable from their social environment and relationships. As Johnstone says, what is classed as mental illness can be a person's attempt to be protected, valued or to find their place.[34]

'Our behaviour is an intelligible response to our circumstance, history, belief systems and bodily capacities,' she says.

Johnstone campaigns for dispensing with diagnostic labels and, as an alternative, suggests describing a person's problem in terms of symptoms (e.g. low mood), then asking certain questions. What has happened to you? How did it affect you? What have you done to survive? Instead of looking for a pill, a person should be empowered to look for a reason why they feel as they do. This echoes my sense that my patients' psychosomatic problems are best understood as a maladaptive response to circumstances and that exploring those circumstances presents the best chance of both recovery and preventing relapses. In depression and psychosomatic conditions, the connection of suffering to life events is often very complex and difficult to disentangle, but understanding the connection between your history, how you live your life and how you feel is a surer way of responding differently to future adverse life events than any tablet.

Medical labels also have the effect of impacting identity in unpredictable ways, a consequence that should not be underestimated. When a disorder becomes referred to as a genetically determined, brain development disorder or as a consequence of serotonin deficiency, that risks making it an inextricable part of the person. A person cannot change their essential genetic makeup and while a brain can be trained, a developmental abnormality associated with structural brain changes can't necessarily be repaired. It's permanent.

Sociologists have long studied the effect of labelling on identity. Studies show that when a person is labelled as 'different' or 'other', they sometimes identify with the label and behave differently to conform with it. We exhibit the features of labels given to us. Disorders like neurodiversity come with a set of stereotypes with which people can easily comply. That is not to say that it's a foregone conclusion that every person will take on the features of their label. A person can certainly resist – but when a diagnosis is the only key that opens the door to help, when it validates suffering and attracts a supportive community, why would they?

An 'illness identity' is defined as the set of roles and attitudes people develop about themselves in relation to their illness. Identifying very strongly with a diagnosis has been shown to lead to a more negative health outcome. Even people with bodily disease that is nothing to do with mental health or behaviour have worse outcomes if their illness identity is too central in their lives. A study of people with congenital heart disease found that those for whom the illness dominated their identity had more hospitalisations and more visits to their doctor than others with equally complex and severe heart problems.[35] In other words, the severity of their symptoms and need for hospital treatment correlated better with their strong identification with their diagnostic label than it did with the degree of heart disease. Also, in this group, a strong illness identity was an even bigger predictor of higher medical needs than overt depression and anxiety.

Being part of a group that is centred around illness can also impact outcome.[36] Groups of people who are defined by a mental health diagnosis develop norms of behaviour that can lead to their symptoms and difficulties being exaggerated. This happens at an unconscious level. The neurodiverse community has a high level of group identification. I have spoken to many people who report that learning they were neurodiverse was like finding their tribe. That is wonderful on some levels. It gives people great peace of mind to learn they are not alone. But when a medical problem is at the centre of personal and group identity, it can also threaten wellbeing. There are multiple layers to this. Those who have a strong illness identity can be perceived as less capable by others outside the tribe, which adds to a person's negative self-concept. A person who believes they are incapable behaves as if they are incapable, which provokes others to treat them as if they are incapable, and so the cycle feeds back into itself. There is also the fact that being part of the group may be predicated on being sick and getting better could affect those relationships.

There is a way to break this cycle. Central to recovery is a 'recovery identity'. To get better, an individual must ultimately relinquish their illness identity and develop a new, equally meaningful recovery identity. The journalist Hadley Freeman gives a good example of how this works in her book *Good Girls*, in which she recounts her personal experience of anorexia. Freeman spent her formative teen years in and out of hospital with a life-threatening eating disorder. She was on wards, surrounded by other people with anorexia. There, she inadvertently learned how to be a *better* anorexic, how to keep weight off and how to resist the hospital system. She remained seriously ill until she realised that the only way she could move forward to health was by visualising a life that did not have anorexia at its centre. Recovery only started when she caught a glimpse of what recovery could look like. Obviously, when an illness identity exacerbates symptoms it happens at an unconscious level. Nobody 'wants' to be sick and recovery can be very arduous.

While illness support groups are very beneficial in some ways, they can also be detrimental if they do not have a strong recovery identity built into them. In March 2020, Professor Paul Garner, a UK-based senior medical doctor and researcher, caught Covid-19 and was shocked to find himself severely fatigued weeks after the acute infection seemed to have passed.[37] His initial infection had been mild but the aftermath left him feeling like he had been 'clubbed over the head with a cricket bat'. At points, he felt he was dying. He was utterly exhausted with multiple symptoms that he did not find easy to explain. He had a new symptom every day: muggy head, upset stomach, tinnitus, pins and needles, breathlessness, dizziness. In a blog for the *British Medical Journal*, he described his illness like 'an advent calendar, every day there was a surprise, something new'.[38]

Garner is an infectious disease specialist. He expected that he, of all people, should be able to explain what was happening to his own body, but found he couldn't. As he tried and failed to recover, Garner came to think of his illness as unprecedented. He wondered if the virus had triggered some novel immunological disorder that did not exist in medical textbooks. So he turned to the internet for answers and found he was not alone. In long Covid support groups, there were lots of people who shared his exact experience. Marathon runners who could no longer walk after mild Covid. Through the long Covid groups, he found his way to communities of people who had developed chronic fatigue syndromes after other infections. Many of these people had been ill for decades. Their experiences also resonated with Garner's. His original expectation, based on his medical knowledge, was that he should get steadily better with time, recuperation and gentle increases in activity. But that was not marrying with his unfolding reality. A ten-minute bike ride taken on a good day had provoked a three-day relapse. So he decided to learn from the narratives of those who had been dealing with non-recovery for much longer than him. They recommended pacing, working within the limits of his energy levels rather than trying to exercise his way out of the situation. He took advice from a friend: 'stop trying to dominate the virus; accommodate it.' Garner learned to do less. That brought him to a baseline in which he was not getting worse – but he was not getting better either.

That could have been where Garner's story ended. By September 2020, he had improved but he was no longer recovering further. However, Garner then started searching beyond the non-recovery stories, for those who had more positive outcomes. That was how he found Recovery Norway, a group of people who once had chronic fatigue syndrome but had beaten it. The group gave him a recovery mentor as well as another perspective, and, crucially, a recovery identity. He realised that while pacing had helped him at the start, he had then become obsessed with it. As he described in his blog, he had started to unconsciously monitor signals from his body until he became paralysed with fear.[39] He believed long Covid was a metabolic disease that had damaged his mitochondria, but the Norway group made him think differently about his predicament. He didn't doubt that the virus had triggered the fatigue in the first instance but felt he had later become caught in a vicious cycle of illness driven by his fear. Viruses cause fatigue in order to make people rest, which promotes recovery. People with a viral infection should rest until the acute infection has passed, which is usually within a few days or a couple of weeks. But, in Garner's case, his later recovery had gone awry because he inadvertently conditioned his body to stay tired. Garner realised he had to retrain his brain to react differently to the fatigue if he was to get better.

As he writes, 'I suddenly believed I would recover completely. I stopped my constant monitoring of symptoms. I avoided reading stories about illness and discussing symptoms, research or treatments by dropping off the Facebook groups with other patients. I spent time seeking joy, happiness, humour, laughter, and overcame my fear of exercise.'

By the end of 2020 he had made a full recovery.

ADHD used to have a recovery identity. In the 1960s and 70s, the DSM described it as a condition that went away in adolescence. By the 1990s, it was recognised that the symptoms did not always disappear completely but did lessen as people got older. Some studies found remission in up to 60% of people.[40] [41] Severe ADHD lessened but often persisted while people with mild ADHD could expect a chance of full recovery. But ADHD is slowly being incorporated into the identities of many young people. Some support groups discourage the attempt to overcome ADHD traits. People are told to unmask and to share their ADHD selves with others. But learning to control our moods, behaviour and impulses is part of growing up, whether one has ADHD or not. We all become more socially competent, gain focus and are better able to cope through practice. Encouraging young people to do otherwise may be well intentioned but is potentially setting them up for non-recovery. The rise in subtler ADHD presentations in adults may also undermine a young person's expectation that their difficulties will disappear in time. A growing population of adults have incorporated ADHD into their self-concept. When a medical problem is part of a person's identity, it becomes inescapable.

*

Both the scientific and public discourse around those conditions referred to as neurodivergent have become confused. Doctors and scientists are medicalising and biologising at the exact same time that support groups and some sufferers, those at the mild end of the various spectrums, are asking to be allowed to be their most authentic neurodivergent selves. The medical community is keen to respect the view of neurodivergent people so it is also slowly changing the language of the neurodevelopmental problems to remove words like 'disorder'. Some subgroups of neurodivergent people welcome this but are still perversely reliant on the medical community for psychological support and medication, and on society for special accommodations.

The contradictions are somewhat set up by the way the term 'neurodiverse' has come to be used. It is a non-medical term that encompasses a range of conditions, including depression, ADHD, autism, dyslexia, dyspraxia and Tourette's syndrome. There is certainly a lot that is positive to be said about the use of the term to remind us that all our brains are different and that, as a consequence, we perceive the world and function differently. Our moods, concentration levels, social skills, creativity, all our personality traits and abilities lie on a continuum, affording us different strengths and weaknesses. There is no right or wrong way to behave or feel and all ways of being should be embraced.

However, the problem is that the term is not used in that way. Instead, it is used as a counterpoint to the term 'neurotypical'. A neurotypical person is said to organise their thoughts and behave in a 'typical' way. In explaining this, one ADHD charity website says that neurotypical people can navigate 'socially complex situations with ease'. Another charity website supporting neurodiverse people acknowledges that neurotypical people have social difficulties but says they can overcome them fairly easily. A very common explanation for the difference between neurodivergent and neurotypical people is that neurotypicals were born with an inbuilt handbook for life, which gives them an innate sense of social rules. Neurodivergent people did not receive the handbook and, therefore, have to work much harder to learn and fit in. The division of people into typical and diverse immediately contradicts the sensible statement that we are all different. It brings to mind George Orwell's *Animal Farm*, where the pigs decide that 'all animals are equal but some animals are more equal than others'.

Neurodivergent conditions are still finding their place but what is clear is that people with that label are suffering. Their hardship is real or they would not have sought a diagnosis. Anna has struggled socially and in her work for a very long time. She's had a lot of internal pain and needs support. But the question inevitably arises as to whether the benefits to those diagnosed with ADHD outweigh any drawbacks. If they do not, then this is another example of overdiagnosis – the diagnosis might be right but it is not beneficial.

Unfortunately, ADHD does not stand up well to an overdiagnosis test. It has become standard practice to make accommodations for students with ADHD in school and university, and yet, there is a worrying lack of experimental studies to support their use. The concern is, as always, at the mild end of the spectrum. People with severe ADHD and autism have always and will always need extra support. Special accommodations at school, such as one-to-one support, that allow people with severe ADHD and autism access to education, are essential. They benefit not only the neurodivergent individuals but also the general school population, who needs to learn about the diversity of people. Severely autistic Elijah, who you met in chapter three, was non-verbal when he attended mainstream school. His family are grateful for the support he got that facilitated that. He grew accustomed to being around groups of children and they grew accustomed to him. He wasn't much interested in friends but was able to mix with people he would not have met in other circumstances. He learned at his own speed in parallel to his classmates, courtesy of a personal tutor.

The issue with overdiagnosis is always around the benefit of the diagnosis and value of special accommodations to those mildly affected. The research that has looked at this issue has not produced very optimistic results. One Canadian study compared university students with ADHD, some of whom received no special accommodations and others who were given help, such as extended time for tests and separate test-taking rooms.[42] The students in receipt of accommodations perceived the extra support as helpful but there was no measurable benefit in terms of academic performance. Several studies have looked at school-aged children and students in higher education who were given various allowances, including a reduced distraction environment, use of a calculator, more frequent breaks and oral presentation of written information.[43] [44] [45] Again, those people with ADHD who received this extra support performed no better than pupils with ADHD who did not. One study found that children screened for ADHD in school did not benefit but raised the concern that children were harmed by being labelled at such a young age.[46] Similarly special accommodations made to support adults with ADHD have yet to be proven to make a useful difference to their lives.[47]

A big problem with these sorts of accommodations is that they promise better school or work performance but don't deliver. They risk creating the expectation that extra support is essential. If a student believes they need that support to do well, they may continue to think they need it going forward in life when actually they could have done just the same without it. They will eventually meet a situation in which no special accommodations are available and when that happens, a history of assuming they require extra help could set them up for failure.

Stimulants are not first-line treatment for ADHD in children. They are reserved for when behavioural treatments and support are not enough. The short-term efficacy of stimulants in children is considered to be well established in reducing symptoms such as hyperactivity. They improve concentration and teachers perceive better behaviour in children taking them. However, most studies have only followed the children for a short period of time. What is less clear is whether or not that symptom reduction translates to something more meaningful in the long term, like a better quality of life or academic performance.[48] [49] [50] [51] Stimulant medication only reduces symptoms while you are taking them. For a more lasting treatment effect, behavioural interventions are needed. Again, children with severe ADHD are the ones who stand to gain from stimulant medication because their disability is such that medication has a more palpable effect on their ability to concentrate on tasks such as reading so they have time to learn. Stimulants help them to remain in education. But at the mild end, there is doubt that stimulant use has enough of an impact to offset either the negative effects of labelling or the low expectation that can be inferred from a mental health diagnosis.[52]

It is only anecdotal, but the lack of palpable life improvements for people diagnosed with either ADHD or autism is something that concerned me many times as I interviewed people for this book. Of the scores of people I spoke to, all of whom were adults, all perceived their lives to be better off thanks to a diagnosis. Every person welcomed the diagnosis into their lives. But almost all had left their job, dropped out of education and lost many old friends. Several were housebound. The same can be said about those I spoke to with chronic Lyme disease and long Covid. I saw a worrying gap between the perceived benefit of being diagnosed and any actual improvements in quality of life. In most conversations I was left wondering how long the positive impact of validation would last.

Although, even as I say this, I realise I am measuring their life improvements by my own personal idea of what successful treatment should look like. I expect most medical labels to lead to improved health with fewer symptoms, and that should be followed by doors opening onto new opportunities and an easier progression through social life and hopefully work life too. The people I talked to each got great psychological relief from their diagnosis and were sure it had made their lives better. So they must have seen improvement by *their* measure of what improvement looks like, not mine. Perhaps what they actually needed from a diagnosis was permission to do less in a world that only values very particular types of success. For some, medical diagnosis is a means to take the pressure off so they no longer feel forced to continue chasing after an overly idealistic social and work life.

But they were all adults. They had had time to figure out what they were good at, what they could overcome and what type of life they wanted. Telling a child with difficulties in the mild range that they have a neurodevelopmentally abnormal brain is different, because giving them that identity might mean they never get the chance to learn their strengths and to challenge their perceived weaknesses.

6
Syndrome Without a Name

When Hana was born, she seemed healthy but her mother, Uma, wasn't so sure. Hana was the fourth of four children and Uma sensed she was different to the others. She didn't seem to develop at the same rate as her older siblings. Health visitors tried to reassure the family that she would catch up given time. So they waited for that to happen but all they saw was that Hana didn't crawl, couldn't support herself to sit up and didn't babble like other children.

It took a year for Hana's parents to persuade medical professionals to see what they were seeing. By then, Hana had fallen far enough behind her peers to make it hard for others to deny the problem. Hana was duly subjected to the standard range of metabolic and genetic tests used to try to explain slow development in a child. Whatever tests were available in 2012, she had. Everything came back all clear. Hana was smaller than her peers. She walked late and talked late. She was slow to toilet train. She was delayed in all areas of development. Doctors agreed that Hana had a problem but there was no immediate explanation why.

'They called it global developmental delay. Or syndrome without a name,' Uma explained.

For a very long time, lots of children who did not develop at a normal rate didn't necessarily get a specific diagnosis because there was not enough knowledge to explain every developmental problem. Instead, they were given descriptive labels in place of a diagnosis. In Hana's case, global developmental delay or SWAN (syndrome without a name). It was hard for the family to live with such a non-diagnosis. Recognisable disease names make it easy to negotiate with medical professionals, social services and schools. Not having a shorthand label that encapsulated Hana's specific difficulties meant Uma had to constantly describe her daughter's difficulties to others so they could understand her needs.

Then, in 2016, when Hana was four years old and it seemed no answers would ever be forthcoming, her paediatrician offered the family a part in the 100,000 Genomes Project and they agreed. Launched in 2012, this British initiative set about sequencing 100,000 whole human genomes, the entire set of human DNA, in the hope of offering a diagnosis to people with certain rare diseases.[1] In this project, people with a range of unexplained medical problems, like developmental delay, cancers and epilepsy, were offered the chance to participate. To help make sense of the findings, parents were also tested. The project was a huge success. It found a diagnosis for 18.5% of those recruited.

'We didn't hear anything until three years later,' Uma explained. 'It had been so long since we gave the samples that I'd given up hope. I forgot all about it.'

Uma was equal parts excited and terrified when she got the call to return to the hospital to discuss the results. Her biggest fear was that nothing had been found and they would be back to where they started. But instead, the family got good news, of a sort. A genetic abnormality *had* been identified to explain Hana's difficulties, albeit an exceptionally unusual one – Okur-Chung neurodevelopmental syndrome (OCNS). This neurodegenerative disease is caused by a variant in the

CSNK2A1 gene on chromosome 20. When it is healthy, this gene is responsible for the production of a protein called casein kinase 2 which is crucial for normal development. People with OCNS have motor and language delay and difficulty feeding. They also have a characteristic facial appearance with an upturned nose and wide nasal bridge. Many have sleep problems and some also have seizures. At the time that Hana was diagnosed, only sixty people in the world had ever been give this diagnosis, which is why no doctor thought of it before the genetic tests came back. As of yet, there is no treatment for OCNS.

Once again, the purpose of a diagnosis is to provide an explanation; to indicate the best treatment; for prognostication; to identify care pathways; to help identify support groups and others affected by the same problem. A diagnosis with only sixty sufferers worldwide, for which there is no treatment, doesn't fulfil much of that promise.

'We were thrilled to get a diagnosis but then upset to discover it was so rare,' Uma told me. 'We wanted to know what the future holds for Hana but the diagnosis couldn't even tell us that.'

Uma immediately started researching and discovered a support group in the US for people with the condition. More than anything, she wanted to know what sort of a life a person with OCNS could have as they grew into adulthood. The answer was far from clear because the disorder has a broad spectrum. Some sufferers are well enough to work, albeit in low-skilled jobs. Some of those affected had children, although all needed support to allow them that degree of normal life. Others, Uma learned, were severely disabled, unable to do anything for themselves.

'It was a relief to look at Hana and realise how much worse things could have been for her,' Uma said.

Although Hana's development was slow, she has made progress at her own rate. Aged thirteen, she attends mainstream school. She needs a lot of support to do so and her learning is far behind her classmates. She reads and writes at a six-year-old level. She still struggles to hold a pencil and to cut up her own food. But there are lots of things she loves to do. She's a keen trampoliner who enjoys taekwondo and loves Barbie, swing sets and cuddles. She is also obsessed with soothers and can hold four in her mouth simultaneously.

'She's a little bit like a celebrity at school,' Uma told me. 'When she walks down the corridor everybody says hello to her.'

Hana has been in the same school since the beginning and her classmates are very understanding of her difficulties. She knows she is different to them, but not in a bad way.

The family aren't quite sure how things will develop for Hana but they value having people with whom to share their journey. Everything that Hana struggles to do has been explained by the diagnosis. The prognosis is still somewhat unclear but it has been good for the family to see that people with the disorder do not usually show any deterioration over time and seem to have a normal lifespan. Hana and her family are clear that, for all its limitations, life with an ultra-rare diagnosis beats no diagnosis hands down.

⋆

It took thirteen years to sequence the first whole human genome in an initiative called the Human Genome Project, a global scientific effort that began in 1990 and was completed in 2003, at the cost of billions of dollars. A mere two years later, in 2005, next generation sequencing (NGS) became widely available. This is a technique that allows rapid sequencing of multiple DNA strands simultaneously. Thanks to NGS, full human genome analysis can now potentially be completed in one day for under $999. In 2012, the 100,000 Genomes Project harnessed NGS for clinical use, turning it into a diagnostic tool for individuals. As a neurologist, I work with large numbers of people with difficult to explain neurodevelopmental problems and the project gave several of my patients a diagnosis after decades without one. It has given great psychological comfort to many families and has also been a useful guide to treatment in some.

Before these innovations, a genetics lab could only analyse small sections of DNA and, because processes were so slow, only did a limited number of analyses per week, meaning genetic tests were often rationed. They were mostly done for people in whom there was already quite a strong clinical suspicion that they had one of a specific range of known genetic conditions. Rapid whole genome sequencing means comprehensive genetic testing is now widely available and can be carried out more speculatively than previous genetic testing. It can be done for children in whom the diagnosis is just not clear. The problem is that a shot in the dark like this can produce confusing results.

Henry is Jenna's only child. He was something of a surprise. She had recently split with her partner when she learned she was pregnant. It was far from planned but, after a day or so of panic, Jenna realised she was thrilled to be having a baby. Henry duly arrived at full term with no apparent health problems.

'I wasn't the one who thought he had a problem,' Jenna told me. 'I didn't know what to expect from a baby and nothing very big happened so I didn't spot it. There were little things that other people noticed. He didn't put on weight very well. At one point, the health visitor advised me to take him to the doctor but the doctor didn't do much. And lots of mums had similar stories.'

At first, all the questions about Henry's development came from a health visitor. Later, Jenna developed her own concerns, in part provoked by the health visitor's questioning. She was worried that his movements were jerky, but her mum assured her that was normal for a small baby. Henry wasn't really very different to other children in the first few months, except in subtle ways. He was slow to respond to sounds and was an exceptionally quiet child who didn't cry very much but didn't make happy noises either. He was referred for a hearing test but by the time that was done, he had started to attend to Jenna's voice and to make cooing noises.

'His mouth hung open a lot of the time and one of the mums in the mother and baby group commented on that, which really upset me. My sister made the point that I didn't really know that much about his dad's family so Henry probably just looked like them. She was trying to make me feel better but all she did was make me worry that Henry had inherited some awful disease and I hadn't even done due diligence by finding out more about my child's family history.'

At six months old, Henry still couldn't support his head and seemed unusually floppy. Jenna wasn't sure how worried she should be so finally asked to see a paediatrician. More than anything, she was hoping for reassurance.

'The paediatrician pointed out things that I hadn't noticed. Henry's ears are too flat, apparently. They suggested a genetic test and of course I agreed because I thought that, at worst, I would get a diagnosis and then I'd know what I was dealing with. Or I'd get the all-clear and I could stop worrying. I didn't know there was an in between.'

Henry's test revealed a novel genetic variant. All the other tests were normal.

'I was told that his genetic test showed a possible abnormality, but the paediatrician didn't seem to know what it meant,' Jenna said.

Henry was found to have a variant in a gene sometimes associated with developmental delay and learning difficulty, but

Henry's specific variant had never been recorded before. The doctors thought it *could* be the cause of Henry's slow development but, because it was new, they couldn't say for sure.

Henry and Jenna were referred to a genetic counsellor to discuss what this finding might mean for Henry.

'I promised myself I wouldn't google while I waited to see the genetics counsellor but of course I did. I didn't know what I was looking for so I only managed to worry myself even more.'

When the conversation with the geneticist finally happened, it gave Jenna some hope but also confused her. She was simultaneously told that Henry's variant might indicate a medical problem, but then again it might be nothing. Henry's slow development was so marginally outside the normal range that he still had plenty of time to catch up with his peers. His flat ears might be normal. Then again, the variant might cause learning or developmental problems that would become more obvious as Henry got older. Only time would sort this out.

The geneticist referred to the result as a 'variant of uncertain significance'. This is a genetic variant for which there is insufficient evidence to prove one way or another whether it is benign or pathogenic. In other words, Henry's variant was too unprecedented to be useful in predicting his future health.

Not every variant causes disease. Far from it. More than 72,000 unique variants have been found in the BRCA genes but only around 4,900 of those are associated with a high risk of cancer.[2] Time was needed for Henry's gene to be found in other people so that its significance could be measured. If those people proved to have developmental problems, then the variant would be considered pathogenic and might explain Henry's slow development. But if they were healthy, the variant could be dismissed as benign, meaning it had been a red herring all along. Nobody knew how long that sorting-out process would take.

Jenna and Henry would have to be patient as the story of his variant unfolded. Which could take decades.

When an adult is offered genetic testing, they consent for themselves and have the life experience to understand how an abnormal test might impact them. Tests for Huntington's disease or BRCA come with a long history of research that makes sense of the results, for most people at least. But a variant of uncertain significance is a sort of non-result. In this case, given to a child who might have it hanging over his childhood, perhaps for no good reason.

'I was very upset,' Jenna told me. 'I went to the geneticist expecting an answer and was essentially told that nobody knew. I understand why the paediatrician did the test but I was also a bit angry at the false promise.'

Jenna didn't regret agreeing to the test. She took the attitude, what's done is done. She just wished that she understood that such equivocal results existed so she could have prepared for it.

Later, Jenna and her ex-partner, Henry's father, each had genetic tests to determine if Henry had inherited the variant of uncertain significance from one of them. If one of them had it too, it would go a long way to reassure everybody that Henry might be perfectly healthy. But their tests came up clear, which meant the variant must have been *de novo* – meaning appearing in Henry for the first time rather than being inherited from a parent.

'I've had to fight to forget the genetic test result,' Jenna told me. 'That's what the genetic counsellor said I should do. She said I shouldn't waste time worrying about it. That's easier said than done. Every time Henry takes a spill or cries for no reason, every bug he picks up, it comes to mind.'

Henry is now two years old.

'Is he catching up with his peers?' I asked.

'Henry does everything a little bit late. Walking, talking. He makes you think he won't ever do something you want him to do and then he does it. When I'm worried, I remind myself that my sister walked at eleven months but I didn't start until I was fourteen months. Children develop differently. And he's a very loving child and has a very great sense of fun. That's more important than his word count aged two and a half.'

'Is there follow-up with the geneticist or the paediatrician for this?' I asked.

'No!' Jenna laughed. 'They dropped the result, saw Henry a couple of more times, told me to watch and wait, and that was that. I know they would see Henry again if I asked, but I could tell I was wasting their time. There was nothing else they could say or do.'

*

The genome is a list of letters that feels as if it could be read like a book, but it's a book in a language that is so new that only a small number of words have been deciphered. And, like any language, even those deciphered words could have multiple meanings. The impressive technology involved in genetic testing belies the fact that we have only just garnered the ability to read the whole genome and have barely started on the road to decoding it fully. Genetic results can look like precise formulae but the distillation of that to something that has clinical meaning to the patient is considerably less precise.

Although we have talked a lot about rare variants, variants themselves are not rare. Your genome contains millions of variants, as does mine. Fortunately, only a very small percentage of variants are likely to be associated with disease. Just like with Lyme disease, autism and cancer genes, we are back to thinking about the importance of the clinical context in making a diagnosis. Once again, test results mean nothing without the patient's story that goes with them. What that means in the case of genetic tests is that the doctor caring for the patient gives the geneticist a detailed description of their patient's problem (phenotype). Although the whole genome is sequenced the whole genome is not analysed because that would return a dizzying array of uninterpretable variants. Instead, the geneticist uses the patient's story to focus their search on a specific range of genes previously associated with that phenotype. The quality of the result the geneticist comes up with will depend heavily on the quality of the story provided by the referring doctor.

Take the example of a child under investigation for an unexplained learning disability. Their genome will be analysed by looking at genes already known to be associated with learning problems. A search like that will turn up thousands of variants in the first instance. Then a filtering process begins. Common variants seen in lots of healthy people and not known to cause learning problems can be filtered out. This filtering process contains a lot of subjectivity and imperfect algorithms. A genetic screen doesn't give a positive or negative result but rather a range of potential abnormalities that are then subject to interpretation by computer and human combined.

The reliability of a genetic diagnosis is at its best if there is a very specific and even rare phenotype. For example, imagine I refer a child for genetic testing who has developmental delay, seizures, flapping movements of the hands, an oversized tongue, wide-set eyes, stiff jerky movements and diabetes. With that description, I have provided the geneticist with the sort of detailed phenotype that allows them to filter the inconsequential variants more easily and focus in on genes that, when diseased, are likely to produce this particular constellation of characteristics. However, if the only phenotype I provide is one of mild developmental delay, it would be very hard to know which of thousands of incidental variants to dismiss. Developmental delay is both too vague and too common to allow for an easy and meaningful genetic analysis.

These principles apply to most tests and many diagnoses, which is precisely why the diagnosis a person is given can vary depending on how many bits of the jigsaw are gathered and how successfully they are put together. Going back to the example of MRI scans we explored in a previous chapter, imagine a person with fatigue was referred for a whole-body MRI scan. How would the radiologist possibly know where to look on the scan for the explanation with a medical history as vague as that? How would they know if some small, innocent-looking cyst on the lung was something or nothing, if they didn't have more specific symptoms with which to work? The quality of most test results reflect the quality of information the clinician gives to the person reporting the test.

Genetics results are impressive in their newness and their complexity. People expect a higher standard of result from genetic analysis than they do from a scan or simpler blood test. But genetics tests are subject to the same errors and difficulties as most other tests. Interpreting their meaning cannot be done by a computer alone. The technology might be cutting edge, but diagnosis still rests as much as it ever did on the quality of the doctor.

*

Jenna doesn't know what to tell people when they ask after Henry's wellbeing. She doesn't know what to write on insurance forms. She doesn't know whether to advise prospective schools that he might have a learning problem. Because he also might not. When she looks at the medical record that their family doctor holds for Henry, his gene variant is listed as a diagnosis with such certainty that it feels like something that needs to be declared and which will affect the rest of his life. It's easy to forget that the geneticist didn't know what it meant. In my clinical practice, I often see variants of uncertain significance listed as a diagnosis in medical notes. They insinuate themselves into a patient's record where they take root. Even if they are later proven to be meaningless, they may still never disappear from the diagnosis list.

With Henry in mind, I asked Professor Anneke Lucassen if she thought it might be a better practice for doctors to stop reporting variants of uncertain significance to patients and families, if nobody actually knows what they mean. Was more harm done than good when a paediatrician gave Jenna Henry's result?

'I suppose the question most people would ask is: who are you to want to withhold that information from people?!' she laughed, wryly. 'People have a right to access their data.'

As a clinician, I admit to an inclination to try to protect my patients from uncertain information that I think will only worry them. I have seen the harm that medical labels do, how they can create a sick role, how they can limit people's dreams, and that worries me. People enact the features of a disease just through anxiety about that disease, through searching their bodies for symptoms and through expectations, so what effect will a genetic diagnosis have on a child's identity, behaviour and view of themselves going forward? Will Jenna be able to resist communicating her worry to Henry, potentially eroding his self-belief without meaning to?

Henry also brought to mind the HD community's preference not to know about a future that cannot be changed. And Stephanie, my epilepsy patient who was grateful that she didn't find out about her KCNA1 variant until middle age, which had allowed her to enjoy a free life. Withholding information from a patient because it is ambiguous is viewed as condescending.

I know that, but still struggle with finding the balance between knowing which of my patients will find uncertain test results tolerable and who will take them on like an albatross around their neck.[3]

I was glad to realise Professor Lucassen did not entirely disagree with me because she added, 'Although, I personally think the paternalistic argument comes from the very deterministic point of view – people think that a genetic code says something very clearly when really it doesn't.'

Genetic determinism is the belief that genetic coding is much more important to a person's health and development than external factors. Genetic determinism gives people no credit for how behaviour and environment impacts gene expression. It gives no credit to epigenetics. As Professor Lucassen had advised me before, a person's postcode at birth can potentially predict risk of future disease just as well as the genetic code.

I asked Professor Lucassen if she thought there was a way to report variants of uncertain significance to patients and parents that might leave them less in limbo.

'You keep asking the wrong question,' she advised me. 'The right question should be: is it even a *thing* to report?'

Clearly, when doctors find information in a genetic test that seems to explain a person's specific symptoms, even if it does not come with treatment or prognosis, that information should be passed on to the patient or family, even if its meaning is somewhat ambiguous. But, perhaps, if a variant is found in a gene that may or may not be relevant because nobody actually knows what it means, then the proper answer is to find ways and systems to understand that result rather than automatically worrying the family. When science knows a bit more, then the patient can be contacted with that finding. In many senses, genetic tests that return uninterpretable results are essentially negative tests that failed to find an answer. But there is a real risk that those non-results are being conflated with something more meaningful than they really are.

In my practice dealing with rare, difficult to explain brain disorders in adults, when a variant of uncertain significance is uncovered, it isn't automatically reported to the patient if it is deemed uninterpretable. Instead, it is my responsibility to keep that variant in mind so I can review the database of variants on a periodic basis, as I wait to see if more information allows it to be moved either into the category of clinically significant or unimportant. I do not need to pass uncertain results to my patients until there is enough information to make them helpful. Our genetic capabilities moved from labs to patients' lives before we had a chance to learn the meaning of everything we find. It is possible, likely even, that lots of people are being given non-information as fact.

More scientific research into the meaning of the genetic code needs to come into play before lives are impacted by uninterpretable findings. Nothing is lost by putting the discovery into the waiting-to-be-deciphered pile to be fed back to the patient later, if it turns out to be useful. The main argument for releasing all results is one of autonomy but is this really best served by handing out results to unprepared families that not a single scientist or doctor understands?

*

The question of whether or not genetic tests are a benefit or drawback to a child's life is rapidly becoming more pertinent. Henry's test was done because he had a possible, albeit vague, medical problem. But, already, genetic testing is being extended to healthy children in the form of screening.

It's a worldwide trend. The UK-based Newborn Genomes Programme, launched in 2022, aims to use rapid whole genome sequencing (WGS) to 'evaluate the utility and feasibility of screening newborns for a large number of childhood-onset rare genetic conditions'.[4] The Guardian study already underway in New York state has the same aim.[5] Both projects hope to sequence the genomes of 100,000 newborns to look for a specific set of rare genetic conditions. Meanwhile, in Victoria, Australia, researchers are preparing to launch a new project called BabyScreen+ that plans to carry out WGS on 1,000 newborns.[6] These projects will test apparently healthy babies in an attempt to predict future serious health conditions so they can be treated at the earliest stage possible or even before the symptoms start. Of course, babies are already screened for multiple specific diseases at birth, but WGS is much more comprehensive screening.

Nobody presumes to know that extended newborn screening is the right course of action. That is part of what the projects underway hope to assess. And there appears to be a public thirst for it. So far, the New York-based Guardian project has an uptake of 75% of those parents asked to participate.

There are obviously a lot of ethical considerations when it comes to genetic screening in babies. Each of these projects has been designed by geneticists with that in mind. Nobody wants to burden a child with a diagnosis of an untreatable condition like Huntington's disease or to tell a child that they have a high risk of dementia or cancer in their very distant future, so these screening programmes will not look for untreatable or adult-onset conditions. Neither will they report variants of uncertain significance. The project groups have narrowed down the conditions being screened to those childhood-onset genetic diseases that have predictable genetics and where early or pre-symptomatic treatment will make a difference to the outcome of the disease. The aim is to empower people to access early intervention to protect their child's future health, not to weigh a child down with an inescapable destiny.

Nevertheless, here, again, a diagnostic test is being turned into a screening test. Like with BRCA variants, the predictive value of a positive test for the conditions being screened has mostly been applied to people either with the typical phenotype of the disease or those with a strong family history of the disease. The newborn screening programmes will treat every pathological variant as if the disease is an inevitability, giving no acknowledgement to the lessons learned from the UK Biobank cohort – this is the database of healthy individuals in which a significant number of older people have been shown to carry pathological variants that *should* have caused disease but *didn't*. The full prevalence of pathological variants in healthy adults who do not have the disease will not be known until further studies on the scale of the UK Biobank are done on bigger and more diverse populations.

Many of the conditions screened for in these newborns can be treated but not necessarily cured. And while some of the available treatments are fairly simple and generally harmless, like vitamin supplementation or dietary changes, others are more invasive. Some of the conditions are treated with infusions of enzymes, gene therapy, toxic drugs and risky interventions like bone marrow and stem cell transplantation. Subjecting a sick child to invasive therapies is hard enough, but what these predictive diagnostic screens ask is whether or not we should do the same to a healthy – as yet – child. There could be a challenging time coming for these children's parents and for paediatricians in the near future. Although, so far at least, the evidence suggests many paediatricians might be OK with that. One recent survey of 238 US-based rare disease specialists found that 87% were in favour of extended newborn screening.[7] But will they feel the same when presented with a perfectly healthy newborn and their terrified parents in receipt of an unprecedented predictive diagnosis?

How much follow-up and monitoring will be required to decide if and when to eventually treat? The health economics of screening for disease means that overdiagnosis is inevitable, with the only question being what amount of overdiagnosis society is prepared to accept.

An interesting thought experiment which focused on cystic fibrosis illustrates the issues with newborn screening. In 2012, the UK National Screening Committee set up a dialogue to elicit the public view on a potential expanded screening programme for cystic fibrosis (CF).[8] CF is a serious condition that causes damage to the lungs and other organs, and which foreshortens lifespan significantly. There's no cure but strategies that include medication to control symptoms and vigilance in treating chest infections can benefit people who are diagnosed early.

The thought experiment asked a focus group to consider whether they preferred standard screening for CF, which inevitably misses some cases, or whole genome sequencing, which would miss no diagnoses but would detect borderline cases, who would then need several years of monitoring to establish if they really had CF or not. Missed cases meant some children would be diagnosed as older children, thus being deprived of early interventions like physiotherapy and antibiotics that might help preserve some lung function. Borderline cases found on WGS meant some children would need regular hospital visits that might later prove unnecessary.

Before taking part in educational workshops designed to dissect the two approaches, the majority of the participants in the focus group expressed a preference for a change from standard screening to WGS. They did not see the harm in a degree of overdiagnosis if it saved some children from a delayed diagnosis. However, after the full educational dialogue had taken place, some of the group changed their minds. In their second vote, the majority of the participants expressed a preference for keeping traditional screening tests over changing to WGS, this time preferring a slight underdiagnosis to a slight overdiagnosis.

This is the sort of information that changed their minds. In the UK, standard screening misses up to ten CF diagnoses per year. A screening programme using WGS would not miss any genuine CF cases but would detect eighty borderline cases per year, most of whom would not prove to have CF. This means that to avoid a delayed diagnosis of CF in ten children, up to eighty children would face monitoring that would later prove unnecessary. Saving some children from a diagnostic odyssey and from delayed diagnosis through screening using WGS means sending other perfectly healthy children on their own diagnostic odyssey that will lead nowhere. Early life is an important time for bonding. Pre-symptomatic diagnosis will medicalise that time for some healthy children, requiring hospital visits and investigations. It will lead to inevitable parental anxiety. The impact of these factors is hard to quantify as yet. To assess what is lost by turning a child into a patient could take decades to discover, if it is even quantifiable.

The Newborn Genome Programme plans to test for 223 conditions, the Guardian study for 250 and BabyScreen+ for 500. While most children will be recruited during pregnancy, some families in the Guardian study have been approached at the bedside of their newly arrived healthy child. There is a real risk of information overload when parents are asked to understand and consent for their child to be assessed for such wide-ranging conditions. How can medical professionals and institutions apply the normal principles of informed consent when testing for hundreds of disorders simultaneously? Is it even possible in such a setting?

Again, I think of one of geneticist Dr Tadros's lessons. People at risk of Huntington's disease go to see her with the sense that testing for HD is the responsible thing to do. But ultimately, most of those people are relieved when she gives them permission *not* to test. To what degree will a fierce need to protect their vulnerable newborn child influence a parent to think that availing of the very latest medicine has to offer must be better than no medicine? Turning away the offer of some cutting-edge science might feel very difficult for parents who want to do the best for their children.

And the child is a passive recipient of this care. An adult undergoing screening has the opportunity to understand the issues. But newborn screening has the special challenge that the person being screened is not the person giving consent. Parents have to give consent on behalf of their children all the time, but a whole genome sequence may have unforeseen implications for that child's entire lifetime. In the 100,000 Genomes Project, each whole gene that was sequenced was first analysed with the express purpose of looking for an explanation for that person's previously unexplained medical problem. So far, the 100,000 Genomes Project data has been accessed by more than 3,600 researchers from 354 institutions in 33 countries. Going forward, that genetic material will be subjected to a great deal of further testing for many other purposes. The newborn samples will likely also be subjected to multiple levels of testing that continues throughout that person's life. The whole genome is on record and it will remain available to the researchers for that child's lifespan. Participants might find themselves contacted at any time should new findings emerge that could be important to their health. Which, of course, could be of huge benefit to them, but one must always keep in mind the potential psychological and social impact of being diagnosed in that way, which is not something medicine is always very minded to do.

David Curtis, honorary professor at the University College London Genetics Institute, has expressed his concerns publicly, saying, 'Once in a very, very, blue moon, you will pick something up – they are very tiny numbers. The value of the data is not necessarily for the individual. It's being sold to research institutions, universities and biopharma companies, and they will pay for that. When you're signing up to get your genome sequenced, you're also signing up to allow these other companies and collaborators and universities to have ongoing access to your health outcomes.'[9]

*

A child does not even have to wait to be born to be subjected to state of the art genetic diagnosis. In 1997, scientists discovered that placental DNA could be detected in maternal blood. Placental DNA is usually identical to foetal DNA, so that opened up the possibility of genetic testing on a foetus by means of nothing more unpleasant than a maternal blood test. This led to the development of non-invasive prenatal testing, or NIPT.

Screening a foetus for health problems is nothing new either. Every woman undergoes a twelve-week ultrasound scan, which is the first point at which potential developmental problems might be detected. Genetic screening, however, is not so routine. It is only done if the standard screening raises some concern. Before NIPT, the only way to carry out a genetic test on a foetus was either through amniocentesis, a procedure in which a sample of amniotic fluid is removed from the womb, or by chorionic villus sampling (CVS), a placental biopsy. Both are unpleasant procedures for the mother and come with a 1 in 100 risk of miscarriage. The results produced by NIPT are less reliable than those with amniocentesis or CVS, but that is offset by the fact that NIPT, being only a blood test, presents no risk at all to the pregnancy or to the mother. NIPT has become gradually available throughout the world since 2011.

Nicole and Tom first learned that there was a one in forty-seven chance that their unborn baby had Down's syndrome at the twelve-week scan. The news was very unexpected. Down's syndrome is more commonly seen in the children of older mothers, but Nicole was only in her twenties. This was her first pregnancy and she was not prepared for this kind of news.

Down's syndrome has long been one of the main conditions subject to prenatal screening. Current standard clinical practice recommends combined screening to take place between week ten and fourteen of pregnancy. This involves an ultrasound to assess the amount of fluid at the back of the foetal neck and a blood test looking for biochemical markers of Down's syndrome in the mother's blood. These tests are not diagnostic; they only assess risk. Anything less than 1 in 150 risk of Down's syndrome is considered a low-chance result – which doesn't mean there is no chance the baby will have Down's syndrome but makes it unlikely. Any risk greater than 1 in 150 is a high-chance result. In the past, the only options for more definitive tests if a woman was given a high-chance result were amniocentesis or chorionic villus sampling. Many parents preferred not to proceed to this level of testing because of the risk of miscarriage. NIPT has added another layer to prenatal screening because it offers less risk than amniocentesis but more accuracy than combined screening. Like combined screening, NIPT is not diagnostic. It does not return a definite yes or no result, but a high-chance or low-chance result. Parents who want a more definite answer must still have amniocentesis or CVS. The NHS Inform website states that if an NIPT returns a high-chance result of Down's syndrome it will be right 91 out of 100 times.[10] Many other sources quote an accuracy of 99%.[11] [12]

Nicole and Tom did not even consider amniocentesis when it was offered to them. They were quoted a 1 in 100 chance of miscarriage. They knew they could never accept that risk.

They did agree to NIPT. Shortly after, Nicole was told in a phone call that her baby had a 95% chance of having Down's syndrome. She recalls crying when she heard the news.

'I feel ashamed now, about ever feeling sad about the fact that I might have a child with Down's,' Nicole said.

Nicole's upset was very short-lived. One conversation with Tom and she adjusted her expectations for her child's future. The couple was quickly back to being excited about welcoming their first child. When Isobel was born, the diagnosis of Down's syndrome was confirmed. She is now four years old and has just started school.

'What's she like?' I asked

'She's such a happy little bean! That's her nickname – Bean.'

Isobel's development has been slow. She was three and a half before she took her first steps. She is small for her age. She cannot yet talk in full sentences.

'Not that it stops her babbling all the time,' Tom said. 'She has her own little language that she thinks everybody understands, so she just babbles away.'

'Do you worry about her future?' I asked.

'She won't ever be academically amazing. We know that,' Nicole said.

'But neither are we!' Tom laughed.

Nicole smiled. 'Sometimes I look at her and I realise she'll never have any of the ordinary worries in life. She'll get to do pretty much what she wants. Pretty sweet life, really!'

Nicole and Tom were two of several people I met to discuss the experience of prenatal screening and of learning they were having a child with Down's syndrome through NIPT. Like most of the parents I met, they were shunted fairly quickly from their twelve-week scan to NIPT without any real sense that they could refuse further tests. But, despite the automatic feel of it all, the couple were glad to have had access to NIPT and grateful to have a prenatal diagnosis. It prepared them for Isobel's birth and gave them a chance to plan.

Ayla and Omar had a more negative experience. Their son Arun also has Down's syndrome. Arun has an older brother with multiple medical problems, including epilepsy and severe autism.

'I told them I didn't want any more screening,' Ayla told me.

Like Nicole, Ayla was told at her twelve-week scan that there was a very high risk that Arun would have Down's syndrome. Terminating her pregnancy because her child had Down's wasn't something Ayla would ever consider, so she refused further screening tests. She was surprised by the response of medical staff. Some seemed confused by her decision and others made it plain they thought it was foolish. One nurse heavily implied that she shouldn't have had the twelve-week scan if she intended to ignore the findings.

'I was treated like I was turning down some life-saving treatment for my child. But it was the opposite. They made me feel like a bad mother.'

Feeling under pressure, with repeated assurances that NIPT was safe, Ayla agreed to it.

'They sat me down to give me the result. It was very formal. They even said it was bad news,' Ayla told me. 'They talked to me as if my baby was some sort of unfolding tragedy. If I didn't know my own mind and I didn't have Omar, I don't know how I would have resisted the pressure to terminate.'

As the pregnancy progressed, Ayla found the staff more supportive but she still dreaded hospital visits because at every one she was reminded that she didn't have to go through with the pregnancy.

'They kept saying he would have high care needs. But I have a child with epilepsy and severe autism and I can assure you, Arun has much lower care needs than his brother.'

Ayla was forty-four when pregnant with Arun. She and Omar already had three children. Medical staff reminded the couple that, as they grew older, a Down's syndrome child would not only be a burden to them but, ultimately, to their other children too. Ayla was offered a termination seven times during her pregnancy with the last offer a week before Arun was born. While most terminations in the UK are restricted to before twenty-four weeks, Down's syndrome babies can be aborted right up until the day of birth.

Arun, who is now twelve, loves football, computer games and eating sweets. And he is cheeky and playful and full of fun.

'He actually reads and writes better than several children in his class,' Ayla told me.

NIPT faces the same challenges in interpreting the true meaning of results that most other tests face. The 99% accuracy figure quoted by many websites naturally leaves most people assuming that a positive result is pretty much a guarantee that the child will have Down's syndrome. But that's not actually true. The accuracy of a test refers to the proportion of *all* tests that are accurate. Since most people don't have Down's syndrome babies, most negative tests will be right. But the 99% accuracy does not hold true for all positive tests. It all depends on the person being tested.

The reliability of tests for Lyme disease depends on the pretest likelihood that the patients have a high chance of having Lyme disease, as determined by whether or not they live in a Lyme disease area and have typical symptoms of Lyme disease. The meaning of a pathological BRCA variant to a woman depends on her family history. Similarly, the implications and interpretation of a positive result on NIPT depends on the pretest likelihood that the baby has Down's syndrome. Despite the accuracy figures quoted, the chance that a positive test is correct in saying the baby has the condition is often considerably less than 99%.

This is how the Nuffield Trust, an independent thinktank, explains the issue. In most age groups, Down's syndrome affects considerably fewer than 1% of all pregnancies. That means that a dummy test that gave everybody a low-chance result would also be 99% accurate.[13] Therefore, the accuracy of the test is not a very meaningful way of expressing the dependability of the result. The important question is: what is the chance that a positive result is a true positive? That is assessed with a statistical measure called a positive predictive value and that makes the reliability of NIPT look quite different.[14]

'Sometimes, NIPT gives a high-chance result when the foetus does not actually have the condition. If you receive a high-chance result for Down's syndrome, there is a one in five (20%) chance that the result is wrong and your foetus does not have the condition,' said Catherine Joynson of the Nuffield Trust.[15]

The pretest probability feeds into the accuracy of the test. In an older woman or any woman who is considered to have a pretest high risk of having a Down's syndrome baby, the likelihood a positive test is a true positive is often high, maybe 80–90%. But in a younger woman or any woman with a low risk of having a Down's syndrome baby, the likelihood that the positive result from NIPT is correct could be much lower than that.[16] One Dutch study reviewed the outcome of pregnancy in 239 women who received an NIPT result that suggested a high risk of Down's syndrome. Nine of those 239 were false positives, meaning the babies did not ultimately transpire to have Down's syndrome. In that study, there were also five false negatives, meaning five babies judged to be at low risk were later found to have Down's syndrome.[17] The bottom line is that the test may be 90–99% accurate for some women, but that depends on who is being tested, so real accuracy has to be assessed for individuals, not just for the group.

Although NIPT is categorically not a diagnostic test and a positive NIPT is usually followed by a recommendation of amniocentesis or CVS, it is still slowly replacing the invasive tests because it is perceived as safe and reliable. The US hospital Mount Sinai West saw the percentage of patients having invasive tests drop from 38% in 2010 to just 2% in 2015, after the introduction of NIPT.[18] That is totally understandable. Few parents want to risk a miscarriage. But it is also potentially a problem if parents think NIPT is more accurate than it really is.

The rate of termination of pregnancy because of a prenatal diagnosis of Down's syndrome is thought to be approaching 90% in those European countries in which terminations are readily available.[19] [20] In the US, between 67% and 85% of pregnancies with a prenatal diagnosis of Down's pregnancies are terminated.[21] The level of disability is different for every individual with Down's syndrome. Some, like Isobel and Arun, can go to mainstream school and live very full lives, comparable to their peers. But others are born with serious cardiac, bowel and brain development problems, and will have much more difficult lives. The severity of the medical problems isn't always known until the child is born. Twenty to thirty per cent of Down's syndrome foetuses will not survive to term. For all these very valid reasons, most women do not feel able to go through with their pregnancy once the high risk or diagnosis of Down's syndrome has been determined. Some parents just cannot bear the thought they might have to watch their child suffer. Many do not feel they have the emotional or financial resources to provide a disabled child with a good life. Most people with Down's syndrome need at least some level of long-term support and, since parents of Down's syndrome children are often older, they worry about who will care for their child when they're gone.

Naturally, the Down's syndrome community worldwide is very worried about how screening is impacting people with

Down's syndrome and their families. They are concerned that the enthusiasm for screening, now made so easy by NIPT, could see Down's syndrome screened out of existence. They are also concerned that misplaced trust in a positive result from NIPT could lead to women terminating healthy pregnancies. That is not to say they do not see the value of NIPT. It is undoubtedly helpful for parents to have time to prepare emotionally and practically for a child with a genetic difference. NIPT also allows an earlier diagnosis for many parents, which avoids the trauma of a late-stage termination. But they are also worried about how a wider roll-out will impact on the lives of people with Down's syndrome and also the future of Down's syndrome. The number of children across the world with Down's syndrome has decreased significantly due to screening. In Iceland, Down's syndrome has almost disappeared.[22]

NIPT's great advantage is that it is quick and easy, and more accurate than traditional combined screening. But it is just that speedy, safe reputation that has become a concern for disabled communities. Amniocentesis is slow and labour intensive, and there is risk involved – but if there is a perverse upside to that, it is that the drawn-out process gives parents time to think. The procedure gives gravity to the decision to have the test. As a blood test that takes five minutes to do, NIPT could become routine for medical staff, until it is done as a matter of course. As a non-invasive safe test, NIPT will be easier to do and therefore harder to refuse. As the most up-to-date type of screening, it could make parents feel as if they are depriving their child by turning it down. The more modern a medical procedure is perceived to be, the more it is associated with superior medical practice. NIPT could propel parents much more quickly down the path to diagnosis and, from there, to a decision about whether to continue with the pregnancy. Genetic prenatal screening becoming routine risks a future in which medicine screens out difference and stigmatises disability. As NIPT becomes usual, it is essential that the seriousness of its purpose is not undermined by its simplicity.

The assumption that more prenatal genetic screening is only for the best immediately raises the questions of what is a life worth living and what counts as disease. That has been a concern of mine throughout writing this book. I have spoken to many people who carry genetic variants for HD and BRCA, to people who have survived cancer multiple times, to people with rare, untreatable genetic disorders and parents of people with rare or poorly understood genetic disorders, to people with lives plagued by mental health issues, to those bedbound for many years with chronic illness. I did not have the space to tell all of their life stories, but I could tell you a handful of amazing or funny or poignant stories about any one of them. But had they been conceived at a different time, some of those people might have been screened out. Does one have to live to a heathy old age to count? Is it even fair to call some of these problems disorders? Are Down's syndrome or Henry's still-mysterious 'variant of uncertain significance' even medical conditions or are they better referred to as a genetic difference?

In the case of Down's syndrome, the justification for screening is often presented as an opportunity to spare the suffering of the child and the parents, and to alleviate the burden on the parents and the health service. But does a person with Down's syndrome suffer or cause suffering so much more than any other person? People with Down's syndrome are more likely to have health conditions like heart disease, dementia and epilepsy, which means higher than average healthcare needs, and the level of disability does vary. However, their needs are not necessarily any more costly than for people with serious multisystem disorders like diabetes. In adulthood, obesity and smoking are a much greater burden to any healthcare system than Down's syndrome. Although screening is not done with the express intent to terminate the pregnancy, it creates the impression that having a child with a genetic difference is a problem that needs to be solved.

I have wondered for a long time why Down's syndrome in particular has always been such a focus for prenatal screening and the only conclusion I can come to is because it is easy to diagnose. The genetic cause for Down's syndrome was discovered in 1959, decades before the gene for HD was identified. Down's syndrome occurs when a person has an extra copy of chromosome 21, referred to as trisomy 21. The three medical conditions for which there is routine prenatal screening are all trisomy disorders – the other two being Patau's and Edwards' syndromes. As conditions caused by an extra chromosome, these three have been easily diagnosable for a long time. The techniques required to detect an extra chromosome do not require sophisticated next generation sequencing. Screening for Down's syndrome has been routinised to the degree that few question the reasoning for it – except, of course, for the people with Down's syndrome and their families who are concerned about how casually it is sometimes done. And that is my concern for NIPT too. It will make extended prenatal diagnosis so easy that parents may soon think it is not a choice and healthcare workers may start to offer it thoughtlessly.

As usual, there is also the problem of commercialisation. Advertising standard agencies have already had to clip the wings of some fee-per-service clinics that quoted the 99% accuracy figures to parents without any of the necessary qualifications.[23] Some – though not all – private clinics have also started offering to use NIPT to test for a wide range of genetic conditions. But the reliability of NIPT is entirely dependent on how common a condition is. The rarer a condition, the less likely the test is accurate. In 2022, the *New York Times* published the results of an investigation that demonstrated positive results were wrong in the region of 85% of the time for rare conditions.[24]

NIPT aside, the commercial sector has also started to offer whole genome sequencing to parents having IVF. It is an upselling service, promising to give anxious parents reassurance that their baby will be healthy. But it is more likely to cause anxiety than assuage it. Since every person has thousands of variants, it has a high chance of finding uninterpretable variants of uncertain significance. The terrified parents of children screened in overseas private clinics have already started to appear in Dr Shereen Tadros's London-based NHS genetics clinic, but she cannot give them the level of reassurance they want.

'Without a phenotype, it's impossible to resolve what a variant means to a foetus or child,' Dr Tadros advised me.

There is also some interest in using prenatal genetic diagnosis to create 'designer babies'. Companies offer to measure the baby's long-term risk of developing medical problems like diabetes, heart problems and hypercholesterolemia, so that prospective parents can choose the embryo that appears to guarantee the best chance of a healthy life. But, once again, early environment matters more than genetic risk to mental health and lifestyle matters more to general health. Not to mention how little is known about how genetic risks are inherited. Who is to say that a low risk of cancer does not come with a high risk of mental health problems or vice versa?

Roisin, BRCA 1 variant carrier, surgery survivor and mother to two young girls, who we met in chapter four, has thought a lot about all of these issues. Roisin was pregnant with her first daughter before she knew the cancer gene was in the family, but she had her second daughter after she learned she had a high-risk BRCA variant. I asked her about the decision to have a child knowing she had this legacy to pass on.

'Other women I know had PGT to make sure their child didn't have the gene,' she told me, referring to the prenatal genetic testing that allows a woman to choose an embryo with a low risk of carrying the disease gene. 'One woman even asked me, "How do you feel knowing you had a child who could have BRCA?"' Roisin smiled. It was a question to which she had given a great deal of thought. '*Of course*, I don't want my daughters to go through what I went through. Of course not. But if they have to, I intend to support them. I also have ulcerative colitis. I have thyroid disease. If they don't have BRCA, they'll have something else. I can't make a perfect child and I don't think anybody can. My children are supposed to be on this planet for a reason. I'm on this planet for a reason.'

Parents want to bring healthy babies into the world. They want to be prepared for every eventuality. But neither NIPT nor whole genome sequencing nor newborn screening will lead to 'perfect babies' with no health problems. Excessive prenatal screening or screening that is done too casually threatens to undermine the gift of diversity seen in nature and the wonder that is to be found in all children. It is sadder still if poorly understood results and non-information are allowed to overshadow the lives of children and disrupt those important early years when parent and child should be enjoying the prospect of any sort of future.

Conclusion

When I meet a patient with a complicated medical history, I often start by asking when they were last perfectly well. Darcie couldn't answer that question. She couldn't remember ever being one hundred per cent well.

Darcie is twenty years old. She's a patient under my care and her story is one that is very similar to many I hear. She was referred to me because she was having regular seizures, but her illness dated back to long before they started. Despite her young age, she already had multiple diagnoses by the time we met. They had been acquired in stages over the previous seven years.

Darcie's mother agreed that Darcie had always been a sickly child, prone to headaches and stomach upsets, particularly at times of stress or upheaval. The family hadn't worried about that too much when she was very young because the bouts of illness were predictable and manageable. But when Darcie turned thirteen things seemed to get worse. Her headaches became more frequent. Sleep had once helped but that stopped working. That was when her parents grew concerned and took her to see a specialist for the first time. A neurologist diagnosed her with migraine. That was diagnosis number one.

Diagnosis two came about eighteen months later. Aged fifteen, Darcie started to complain of aches and pains in her limbs. Nothing very severe but persistent enough to make her parents take her to the doctor. There was some back and forth on the cause, with the family doctor offering reassurance that didn't seem to help. Then, in conversation with another mother, Darcie's mother stumbled across a condition called hypermobile Ehlers-Danlos syndrome (hEDS), which manifests primarily as hypermobile joints and limb pain. Darcie was considered 'bendy' by her family, so her parents took her to a rheumatologist who examined her and agreed that this was her diagnosis. Darcie was referred to a physiotherapist who gave exercises to help strengthen her muscles and stabilise her joints. There was no cure, but there was hope that she would grow out of this problem as she got older.

After the diagnosis of hEDS, Darcie avoided playing contact sports on the advice of her physiotherapist. Hypermobile joints are prone to injury. By doing less, Darcie certainly had less limb pain but she also became very inactive. Then, aged seventeen, after some persuasion from friends, she agreed to a game of netball in sports class at school. It was a hot day. Darcie, who also had an eating disorder and irritable bowel syndrome by then, hadn't eaten lunch. During the game, she experienced a severe dizzy spell and fainted. The school nurse took her blood pressure and found it was low. That led to Darcie's diagnosis of postural orthostatic tachycardia syndrome (PoTS), a condition in which changes in posture provoke a fall in blood pressure that can cause fainting or dizziness. Darcie was referred for a tilt table test, which measured her heart rate and blood pressure in response to different postures, and the diagnosis of PoTS was confirmed. She was given advice on hydration, salt intake and diet to minimise her likelihood of future faints.

Lifestyle changes didn't help Darcie's PoTS symptoms very much and she continued having dizzy spells and occasional faints. The limb pains and headaches got worse too. She started to struggle in school and was assessed to have both autism and attention deficit hyperactivity disorder (ADHD). Her health went slowly downhill until, eventually, a faint led to a convulsion, which became one of many. At first, her PoTS doctor attributed the convulsions to severe faints and gave Darcie a gradual escalation in treatment for PoTS, moving from lifestyle advice to medication. The medication was supposed to help her maintain a normal blood pressure on standing but it didn't make the seizures any better. They gradually increased in number and that was how Darcie came to be referred to my seizure clinic. Her referring doctor listed all her diagnoses at the top of the referral letter. Migraine, hypermobile Ehlers-Danlos syndrome (hEDS), anorexia, irritable bowel syndrome, postural orthostatic tachycardia syndrome (PoTS), autism, ADHD, depression, anxiety.

When Darcie came to see me, she thought she had epilepsy because that is what one doctor in the accident and emergency department had told her. But that is not what her referring neurologist thought. They were sure that the seizures had a psychological cause. This type of seizure has many names, including psychosomatic seizures, dissociative seizures, functional seizures, non-epileptic attacks and, a very long time ago, hysteria. Psychosomatic seizures are not imaginary or deliberate. They are real and disabling. They arise out of unconscious mechanisms outside a person's control. They are as disruptive to life as a brain disease like epilepsy, but they are not caused by a brain disease.

The biggest challenge for a doctor trying to diagnose the cause of their patient's seizures is that seizures are often rare, so the diagnosis has to be based on nothing more than a description. But Darcie's seizures were happening every day, so all I had to do was admit her to hospital, knowing I would see one very quickly. I run a unit in which I video people having seizures while measuring brainwaves, heart rate, blood pressure and oxygen levels.

Darcie had her first collapse and convulsion as the test was being set up. After that, she had several convulsions over the course of several days. She also had two collapses in which she didn't convulse. Both occurred as she tried to stand up from lying in bed. These she referred to as faints, seeing them as distinct from her seizures. Darcie spent most of her time in hospital lying flat in bed. Some of her seizures happened there, but more happened as she took the short walk to the ensuite bathroom. Two nurses were always needed to support her if she got out of bed because her dizziness on standing was so severe that she constantly felt like she would collapse. This had been the same at home, I learned, where she was sometimes bedbound for days at a time.

Every objective measurement we took during Darcie's stay in hospital was normal. Even when Darcie was deeply unconscious, her brainwaves, heart rate and blood pressure were entirely normal, as if she was awake and alert. When Darcie felt dizzy and fainted, her blood pressure and heart rate were also normal. Her heart rate did rise when she stood up, but only very briefly and by a small amount. There was only one explanation for why these normal measures were contradicting Darcie's experience of her body – the seizures and faints were psychosomatic as her referring doctor suspected. Waking brainwaves are only seen in a deeply unconscious person if the cause of the unconsciousness is psychological. Similarly, while Darcie's early faints, those when she was first diagnosed as having PoTS some years before, might have been caused by drops in blood pressure, those in hospital were not, so they were not due to PoTS. The diagnosis of psychosomatic seizures and faints did not mean that Darcie was not very sick. She clearly was. She had barely been out of her own home for a year. The diagnosis simply meant that a psychological process rather than a disease process was making her collapse.

Any generalist doctor, family doctor, neurologist, rheumatologist, orthopaedic doctor, psychiatrist – in fact most doctors who see a large volume of patients – will regularly see young people with various mixes of Darcie's same diagnoses. Other diagnoses I often see in this combination are Tourette's syndrome, dyslexia and dyspraxia, mast cell activation syndrome (MCAS, a supposed immune disorder), and Chiari malformation (a developmental difference in the base of the skull). What most of Darcie's and these other diagnoses have in common is that they each have a severe form that has been recognised for decades. The severe form usually comes with a demonstrable pathology that defines it as a disease. But, in the last twenty to thirty years, changes to disease definitions have allowed these diagnoses to be made in a milder form that overlaps with normal physiology. The mild form has no proven pathology. The prevalence of the severe form of the disease is stable, while the number of sufferers with the milder form has shot up, vastly outstripping the original disorder. None are curable. Once diagnosed, the only treatment is symptomatic and the hope that the symptoms will lessen as the person gets older.

The stories of PoTS and hEDS are very similar to those of autism and ADHD. There are committees that define physical disease, and that change physical disease parameters to make diagnoses more inclusive, just like the committees that produce each new edition of the DSM. Both PoTS and hEDS are defined by drawing a dividing line between normal and abnormal body measurements. In the case of hEDS, the diagnosis depends on an arbitrary measure of joint movement and skin stretchiness. The diagnosis of PoTS depends on how the heart rate reacts to changes in posture.

Ehlers-Danlos syndrome (EDS) is a long-recognised genetically determined connective tissue disorder that can cause significant disability. It has thirteen subtypes of which the 'hypermobile' type (hEDS) is one. Collagen, the building block for bones, cartilage, tendons, skin and blood vessels, is weakened in people with EDS. The result is hyperflexible joints, skin hyperextensibility and tissue fragility, which can lead to joint deformity, skin scarring, easy bruising and, in some forms, risk of serious haemorrhage.

Of the thirteen subtypes of EDS, twelve are associated with objective biochemical changes in connective tissue. Those twelve all have a proven genetic cause. They are diagnosed on the typical clinical signs with the help of a genetic test. Hypermobile EDS (hEDS) is the only category of EDS for which there is no proven pathology and for which no genetic cause has been identified. It is assumed to be related to the other EDS conditions but there is no evidence to prove that is the case. Its chief feature is joint hypermobility and those affected are said to have frequent joint dislocations. There are no tests to help a doctor make the diagnosis of hEDS. It's diagnosed only on clinical features, with the main sign of the disorder being joints with an unusually large range of movement.

The problem with joint mobility as the main diagnostic feature of the condition is that generalised joint hypermobility is normal in lots of people. It's accepted to be common in young people, with one study estimating that it could be present in as many as 20–30% of healthy people in their late teens to mid-twenties.[1] In diagnosing hEDS, a doctor must determine that their patient's joints are *abnormally* hypermobile. That's not an easy call to make. There is a scoring system that assigns values to the range of movement of joints, but since not every person who is hypermobile has EDS, the diagnosis is ultimately very subjective. The biggest difference between a healthy person with naturally hypermobile joints and somebody diagnosed with hEDS is that the latter probably experienced joint pain and went to a doctor for advice.

Although people diagnosed with hEDS are told it is a genetic disorder of connective tissue, these are assumptions for which there is no evidence. The twelve severe subtypes of classical EDS for which there is an established genetic cause are rare, affecting 1 in 20–40,000 people for most subtypes, 1 in 100,000 for one subtype and 1 in a million for the rarest form.[2] Since the concept of hypermobile EDS as a mild subset of EDS was first coined in 1997, it has become the commonest form of the disorder, increasing the overall prevalence of EDS to as high as 1 in 500 by some people's estimate.[3] Between 80% and 90% of people with EDS have hypermobile EDS.[4]

Postural orthostatic tachycardia syndrome (PoTS) manifests primarily as dizziness, palpitations and fainting caused by changes in blood pressure in relation to posture. Low blood pressure and fainting in young people has long been recognised. In the past, people who fainted with some regularity were given lifestyle advice on diet, water intake and sleep. These measures were usually enough to raise their blood pressure and reduce the number of faints until they grew out of the tendency. PoTS was coined as a discrete medical condition to explain frequent fainting in 1993. The treatment advice didn't change, but this was a new disease label to describe people who were prone to postural changes in heart rate and blood pressure that led to fainting. Nearly thirty years later, it is estimated that somewhere between 1 and 3 million young people in the US have the diagnosis.[5] In the UK, approximately 130,000 people are affected. However, these prevalence figures come from before the pandemic. A lot of people with long Covid have also developed PoTS, so the number of people affected is likely to be higher.[6]

Proponents of PoTS call it a complex, multi-system, chronic disorder of the autonomic nervous system, which is the part of the nervous system that is important for blood pressure and heart rate control. Several diseases, like diabetes and Parkinson's, cause an autonomic disorder. Unlike with these, in PoTS there is no demonstrable pathology or proof of a nervous system disorder. The diagnosis is made if a person's heart rate increases by more than 30 beats per minute when they stand. Everybody's heart rate increases when they stand. The difference for a person with PoTS is that the increase is larger and more sustained. The decision to make 30 beats per minute as the cut-off point for normal was arbitrary, since there is no pathology to prove the diagnosis.

PoTS and hEDS and many of Darcie's diagnoses can be controversial in the medical world because the line between the normal and abnormal measure that allows each diagnosis is drawn very faintly. Some fear that these conditions pathologise healthy bodies by drawing attention to common features of young people's physiology that do not need medical attention. Young people have more mobility in their joints than older people. Young people, and young women in particular, have low blood pressure, making them liable to faint. We become less mobile as we get older and our blood pressure rises as we age, which is why these problems were not considered to need a diagnostic label in the past. They were manifestations of a body that was still maturing. Their symptoms were managed with lifestyle changes for the most part and time was the ultimate cure.

But whether one agrees or does not agree that these are true diseases, what is more important is that they do not stand up well to scrutiny if you examine them for the features of overdiagnosis – meaning the symptoms are real, but the diagnosis provides little benefit and could do harm. Take hEDS. The value of telling a young person they have abnormal collagen that causes hypermobile joints is that it leads to medical advice that should stabilise the joints and lead to better long-term joint health. The principal symptoms of hEDS are pain and fatigue. If the diagnosis is genuinely beneficial then the rise in young people being diagnosed with and treated for hEDS should be reflected in some improvement in a long-term health measure – such as less chronic pain, fatigue and joint disease in older people. But that has not happened. There are just as many older people with joint problems *and* a whole new population of young people with joint problems.

Hypermobile EDS was first described in 1997. Between 1990 and 2019, osteoarthritis diagnoses increased by 113%.[7] Osteoarthritis is increasing in 30–44-year-olds as much as it is in older people.[8] Chronic pain is also on the rise, with two fifths of people affected by their forties.[9] Of course, many other more important factors feed into joint health, especially weight, but for all the millions of people treated for hEDS since 1997 who would not have been treated before then, one might hope to see some health gain, but there is none. Similarly, the growing number of people with PoTS has not led to any improvement in any health measure for any group in the thirty years since its introduction.

It is also worth asking, if there were legions of people with PoTS and hEDS who were undiagnosed pre-1990s where are they now? There is no large group of middle-aged and older people with stories that fit with chronic undiagnosed genetic connective tissue and autonomic disorders, who are suffering the after-effects of a missed diagnosis in adolescence. If they grew out of hEDS and PoTS naturally, then is the diagnosis really needed? More than 85% of people with PoTS go into spontaneous remission. Few people with hypermobile EDS experience serious disability. However, there is a small subgroup of people given these diagnoses who respond very negatively to them. People like my patients, like Darcie, who develop severe disability that is unlikely to be caused by the disease process, but rather by the act of being labelled.

One recent publication in the journal *Brain* described PoTS as a 'fear-conditioned hyperadrenergic state when standing'.[10] A hyperadrenergic state means a person is working on adrenaline. The autonomic nervous system activates in response to fear, leading to the production of adrenaline. Most will know this as the fight-or-flight response. Fainting is frightening, especially the first time. A person who has fainted may well worry that they will faint again. Being told that they have a disorder called PoTS, an abnormal heart rate response to standing, could make that person afraid to stand. If a person becomes very cautious about standing and spends a lot of time sitting or lying down, that deconditions the body, meaning that the inactivity makes the body less responsive to changes in posture just through being out of practice. The deconditioned body is more sluggish in response to changes in posture, making a blood pressure fall on standing more likely and increasing the chances of a faint. It's a vicious cycle, in other words. A person faints. Fear activates the autonomic nervous system. Adrenaline is released. That affects blood pressure, heart rate and behaviour. The body becomes less responsive due to underuse. The fainting gets worse. The fear gets worse. The less a person stands and the more frightened they are, the more the cycle feeds back into itself.

New labels that arise out of moving the parameters of 'normal' create new populations of patients in more than one way. Firstly, by pathologising normal and encouraging people to seek medical attention for something they might not have medicalised before. But also because, once a new disorder is described, some people relate to something in that diagnosis and change to fit with its description. This is an unconscious process. It is what philosopher Ian Hacking calls 'making up people', also known as the classification effect.[11] When a person is given a label or diagnosis it inadvertently instructs them on which new signs and symptoms to look out for, and, in searching their body for other typical signs of the disorder, the attention paid to the body and the expectation of symptoms leads them to register bodily changes that they might previously have dismissed. A classification changes people because, outside of their control, they conform to it.

With this in mind, it is interesting to note that the labels of PoTS and hEDS did not exist until the 1990s, but neither were there a large number of people with the typical symptoms of this diagnosis combination, as there are now. They were so rare in fact that I cannot recall meeting a single person with the specific constellation of symptoms that fit with these disorders until fifteen or so years ago. The growing number of young people with PoTS and hEDS is a new population of patients.

*

Fortunately, most people who are told they have PoTS will not end up as disabled as Darcie, but they may still attract a host of other diagnoses like hEDS. People with hEDS are more likely to develop long Covid. People with PoTS and hEDS are more likely to have mild autism and/or ADHD. These controversial disorders for which there is no proven pathology often cluster together.[12] [13] They are a major reason why I wrote this book. Because not enough people are asking how a growing group of young people like Darcie could be so unlucky as to have multiple apparently unrelated diagnoses all of which are so uncertain in origin. How is a neurodevelopmental brain problem like autism linked to hypermobile joints? Or to an autonomic disorder like PoTS? Why do so many people with long Covid develop PoTS? Researchers have tried to link them through biological theories but they are highly speculative and none are upheld with evidence.[14] These disorders simply cannot be linked through a shared pathology. If that was the case, it would follow that people with severe autism and severe ADHD would have the same high rate of hEDS and PoTS as those with mild autism and mild ADHD. They don't. The association between autism, hEDS, ADHD, PoTS, Tourette's, MCAS and Chiari malformations is only at the mild end of the spectrum. The overlap between all these diagnoses only exists in that grey area of diagnoses where it is hard to distinguish normal from abnormal. Those who try to link them through tissue pathology are, in my opinion, at least a little naïve about the psychology of human beings and about how social factors and medical systems influence the spread of new diagnoses.

These diagnoses coexist because people who express distress through physical symptoms tend to do so with lots of different symptoms that attract a range of diagnoses. And because people who worry about their health tend to worry about all aspects of their health. And because once a person is in the hospital system, they will have tests that pick up small differences that add to their concerns and see specialists who will try to appease them with a diagnosis. Once a perceived abnormality is found, doctors are compelled to keep looking, to monitor, to treat. And when a person receives that first diagnosis, they will be directed to illness groups that warn them of all the other medical problems that might be on the horizon. Joining an illness support group can inadvertently make symptoms worse by encouraging a person to pay more attention to their body at a time when less worried attention is what is really needed. Being part of a group identity that centres around illness can get in the way of a recovery identity.

The more I talked to Darcie, the clearer it was that more than just her seizures were psychosomatic. Everyone experiences physical symptoms when distressed, but for her they were particularly severe. Her headaches and stomach upset during times of upheaval spoke to her tendency to embody psychological upset. Once this tendency had led her into the medical system, she was encouraged to pay attention to and medicalise any bodily change. The more she was asked about symptoms, the more she looked for them. In seeking explanations, she received labels in return. All that did was heighten her health anxiety until she was so overwhelmed that she developed convulsions.

I would like to tell you that once I had established that Darcie's convulsions and ongoing faints were psychosomatic, I was able to reinvestigate and undo some of the diagnoses that had gone before. But Darcie's other diagnoses were not considered within my remit. I strongly suspected that all of Darcie's faints had been psychosomatic, not just those I had seen. I also thought that there was a good chance her hypermobile joints were just part of the normal range of how joints look. I could speculate that her difficulties at school might be better explained by her many medical absences rather than through the two neurodevelopmental disorder diagnoses she had been given – autism and ADHD. I thought Darcie had fallen into a trap of medicalisation that had snowballed. But Darcie found it hard enough to have her seizures called psychosomatic, so when I tried to address her other issues, she put me firmly in my place, saying, 'You're not an EDS doctor and you're not a PoTS doctor, so you have no right to look into those.' Darcie had a point. Specialists are all encouraged to stay in their own lane. So all I could do was add my diagnosis of psychosomatic seizures to her already long list of other conditions. And Darcie went home, presumed to have both PoTS *and* psychosomatic faints. Two diagnoses to explain the same blackouts.

*

Most of the disorders in this book exist in moderate and severe forms and always have. People with unusual novel genetic disorders came to medical attention long before genetic analysis was able to advise them of the exact cause. The family members of people with HD waited to discover if they would also be affected, they just had to wait longer to find out. Children and adults suffered severe mental health problems, learning and behavioural difficulties that were obviously outside the normal range, before it was realised that there could be an explanation like autism or ADHD or some other brain disorder. Specific diagnoses and support may not always have existed but, at this end of the various spectrums, the problems were still obvious.

There is no doubt that those at the severe or moderate end of the spectrum of any of the disorders I have discussed stand to gain from a diagnosis. A person with severe depression has a level of impairment that's hard to miss and there is no doubt they need treatment and support. Some could not survive without it. Kanner didn't have to create the concept of autism for people to recognise that a severely autistic child has some pretty significant requirements for psychological, medical and social support. ADD/ADHD entered the DSM in the 1980s, but even before then, there were children with severe attention and hyperactivity difficulties that were recognisable to all. Treatment services and expertise may have been lacking, social sympathies might have been underwhelming, but nobody needed a label to tell them these children had a lot of extra needs. People severely affected by these sorts of disorders benefit from categories and labels because they lead to treatment pathways, expert researchers, medical specialists, support services. Treatment is justified because the 'symptoms' are such that they do not allow that person to function normally in society and that offsets any downsides that come with being classified as sick.

The same applies to the primarily physical health disorders I have written about. Classical EDS caused measurable disability before the genetic variants were found to explain it. Symptomatic cancer plays by completely different rules to very early cancer cells found on screening. Symptomatic cancer is almost always progressive and potentially life-threatening, while only a proportion of early cancer cells found on screening will ever cause a problem. Disease reveals itself even if you don't look for it or name it.

I do not doubt the benefit of diagnosis and treatment to those with obvious disease and mental health, learning or behavioural problems at the moderate to severe ends of the spectrum because the benefits to that group are clear. In this book, I am questioning the value of drawing people with substantially milder versions of all these medical problems into the diagnostic group. The milder a medical problem is, the smaller the impact of any intervention and the greater the risk of harms from treatment and the labelling effect.

Think of it this way. To a person potentially dying of cancer, the horrific side effects of chemotherapy or risks of surgery and radiotherapy are justified by the life-threatening extent of the disease. A person with a few cancer cells that might never grow will be subject to the same risks and side effects of treatment, but they have significantly less to gain. The harms are the same for both groups but the treatment effect is substantially less for the mild group. Similarly, a person with severe autism or severe ADHD who is supported to go to school with a one-to-one tutor or through the use of medication has all to gain and little to lose through labelling because the severity of their disability speaks for itself. A person with symptoms of autism or ADHD that are so subtle or masked as to be hard to spot is very vulnerable to the labelling effect and has substantially less to gain from medication, school accommodations and other types of support.

There is dubious value in extending treatments and diagnoses to milder and milder populations, but that is what is happening in every field of medicine, for multiple different reasons. We really do think that more of a good thing is still a good thing. In part, this is a cultural problem within medicine. We medical professionals and scientists are seduced by our new technical capabilities. We want to know what else they can do. Rapid whole genome sequencing has only been widely available for twenty years. There is so much still to learn from it and we can't learn without using it. Every few years, somebody develops a better, more sensitive scanner or blood test and that must be used if we are to learn its value. It is exciting when there is a major scientific innovation but I'm not sure that we always communicate the learning curve adequately to the public. Drugs are interrogated by double-blind randomised trials. Diagnostic and technical innovations are not.

I also think that doctors are a little seduced by our roles in people's lives as rescuer and comforter. When a person struggling in school or work comes to us and we diagnose them with autism or ADHD, it is not only validation for that person but for us too. They are grateful and we feel their gratitude. A surgeon who operates and reduces a person's risk of cancer from 85% to 5% has had a good day at work. I am *very* pleased with myself when I make a tricky diagnosis.

But we, the public, should not consider ourselves to be innocent bystanders in the age of diagnosis. In researching this book, I was told a great deal more stories than I could include. In those many accounts of illness, I constantly had the sense that more was being asked of medicine and of medical professionals than they could realistically give. People are struggling to live with uncertainty. We want answers. We want our failures explained. We expect too much of ourselves and too much of our children. An expectation of constant good health, success and a smooth transition through life is met by disappointment when it doesn't work out that way. Medical explanations have become the sticking plaster we use to help us manage that disappointment.

And now the public and medical professionals are caught in a *folie à deux* that we are struggling to acknowledge. There are many more questions put to medical professionals than we can actually answer. Worried people come to see us all the time hoping for a coherent explanation for their problem. We feel our patients' needs and are relieved if we have an explanation to give. It may be that what some of those people really wanted was reassurance, but increasingly, the answer seems to come in the form of a label. That is particularly the case when treatment pathways, records systems and insurers need the label to function.

A lot of people I spoke to in writing this book got great comfort from their diagnoses, even if it led to major surgery or the portent of an awful disease, even when it led to no treatment at all. The source of that comfort was not necessarily inherent in the specificities of the diagnosis itself, but in the listening ear of the person who made the diagnosis and the laying on of hands by the nurse or doctor. That was certainly the case for those who had chronic Lyme disease. Those diagnosed with mild ADHD and autism expressed universal relief to be diagnosed, even though most received minimum medical treatment and their struggles at work and in education continued. But the diagnosis let them know they weren't imagining their problem and they weren't alone and that was enough. Those people who learned they had the cancer gene had to go through a lot, but they also had no regrets because they were given the opportunity to do something active to protect their own health. They had seen family members get cancer and they had saved themselves from that. The rare genetic diagnoses in children were more of a mixed blessing that gave some relief to some people but left others with a feeling of anxious uncertainty about their child's future.

I couldn't speak to the doctors who made all these individual diagnoses but I suspect that if I did, they would also be pleased with the outcomes. They had given their patients an explanation and their patients were grateful for that. Lots of satisfied customers.

Not everyone I spoke to had a wonderful medical journey. Not nearly. The women who had risk-reducing surgery all complained that their surgeons didn't seem to appreciate the gravity of what they were going through. People with Lyme disease in all its forms had to see a lot of doctors and have many tests before they got satisfaction. Those with ADHD and autism were often unhappy with the medical language used and the official descriptions of the condition. There were many complaints about long waiting lists. But almost everyone was happy with the doctor who gave them the final diagnosis.

But, even knowing there was value in these diagnoses for all these people, I still have reservations and fear that the phenomenon of diagnosis creep does more harm than good. Society is not all that good at recognising its mistakes until it's too late, especially when it comes to overusing resources. Antibiotics were a life-saving discovery. They did all the good they promised to do. But in their overuse, we have undermined their life-saving effect. It may be that we needed more and better diagnosis to a point, but now we have gone too far. I worry that we have become so impressed by technology-driven, cutting-edge diagnosis that we haven't taken time to verify that the benefits still outweigh the harms. I worry that those diagnoses that feel good in the moment don't always have lasting value. I worry that a diagnosis that explains our struggles or difficulties can sometimes reinforce those difficulties rather than helping us overcome them.

In 2018, scientists used a sham MRI experiment to show the power of suggestion in children with ADHD.[15] Children and their parents were told that the MRI scanner used in the study was completely inert and was being used as a placebo. (Fortunately, placebos work even if a person knows it's a placebo.) While in the scanner, the children were told that they should expect to feel increasingly relaxed, focused and confident. Eight out of nine children experienced a strong reduction in symptoms and reported an improvement in confidence and self-esteem. Children behave and react the way they are expected to behave and react. As do adults.

We need to be more thoughtful before accepting labels that offer very little and which could lower our expectations for ourselves. When faced with a potential diagnosis, I would like to see people – with the help of health professionals and teachers where it applies – carrying out a balancing exercise in which they ask what the treatment will be, what stands to be gained and what stands to be lost. Central to that discussion should be the consideration of the nocebo effect of labelling and the impact of a diagnosis on how a person is perceived by themselves and others.

Society must also take responsibility for itself by recognising that a wellness culture has made us expect a lot from our bodies and our minds. We have such an unrealistic expectation of happiness that sadness, even when it is entirely understandable, has become pathologised. We so expect to achieve our goals that when we don't, we have started to look for a medical explanation to tell us why. We are encouraged to use medical labels in place of words that say more plainly how we feel.

We need to adjust our expectation of constant good health because it's turning us into an ageist society. We undervalue older people so we resist the inevitabilities of old age. Menopause is a natural stage of life that is currently spoken about as if it was a looming catastrophe for every woman. Certainly it can be an awful experience for some, but for others it's a positive experience and for the vast majority of women, it's a neutral one. Sleep is similarly pathologised and catastrophised. As we get older, we sleep less, but, somehow, popular culture in the form of books and podcasts have programmed many of us to believe that anything less than seven hours of blissful sleep might mean something terrible will happen to us. It won't. Sleep is important but the best measure of sleep is how awake you feel during the day. In the context of an ageist society, it makes sense that loss of mental and physical agility be turned into something medical in the hope that a doctor might be able to reverse its course and, if that isn't possible, that signs of ageing might at least be forgiven by the power of a medical label. An expectation of constant good health, graceful ageing and an obedient body and mind has left people unprepared for those ordinary bodily declines that affect us all.

But I worry most about young people. In a culture that expects success and physical perfection, diagnosis has become a means to account for anything less. Success is not achievable every single time for everyone. A culture of telling people they will get there if they just keep trying isn't fair on everybody. Every single person cannot achieve their most desired goals and we could be a great deal kinder to ourselves and our children if we learned to recognise the point at which we are best advised to readjust our expectations. We would be happier if we learned to better see our real strengths. Sometimes the thing you want to be good at isn't the thing you're really good at. Medical diagnosis has been co-opted to help people deal with these frustrations, but I fear that perpetuates failure and sadness rather than allowing it to be processed and left behind.

In writing this book, I met many people who told me that they could not realise the success that they had so often visualised and their diagnosis helped them to manage the disappointment of that. One woman's story stood out. She's an artist who has had some significant successes, but not the success she wanted. She would prefer to be an academic, she told me, but it wasn't her talent. As a teenager, she imagined her adult self quoting poetry off the cuff. The pain of not turning out to be that adult was terrible for her. It stopped her enjoying the success she had. Eventually, a diagnosis of ADHD helped her accept that she could not be equally talented in every way she wanted. The diagnosis gave her some comfort but also caused stagnation in her life. It reinforced her belief in a lesser version of herself. Rather than being defined by a successful art career that others would envy, she became a woman whose life revolved around ADHD and was defined by the things she could not do because she was neurodevelopmentally different. In this way, I worry that a child told they are neurodevelopmentally different will underestimate themselves and limit their future.

I also question the culture in which a diagnosis seems to be needed in order for a person to access help. I would like to think it should be possible to identify children who are struggling and support them without labelling them. A person should not need to be called depressed to seek help from their doctor or talk to a counsellor because they feel low. But for that to change, insurers and some health services would have to make a major adjustment to how they work.

To solve the overdiagnosis epidemic multiple parties need to make changes. We, the public, need to accept the limitations of medicine and contain unrealistic expectations of what a diagnosis can achieve. We need to be kinder to ourselves when we fail and accept our many imperfections. We need to teach our children to work to their strengths rather than using educational accommodations to hide their weaknesses.

But the medical establishment has quite a lot of work to do too. Firstly, healthcare services need to reconsider how resources are distributed. Investment in staff is more valuable than investment in novel technology that we are only just learning to use. Those new machines may look impressive but they're nothing without the diagnostician to distil their confusing results into something sensible and helpful for the patient. Spending money on staff who can provide psychological and social support is more likely to pay dividends than a wide roll-out of an array of new-fangled tests that may very well do nothing more than produce uninterpretable results or diagnose people with diseases that can't be treated. Many medical puzzles are better answered by a series of quality consultations with a single experienced doctor or nurse than by a scan.

Soon, artificial intelligence is likely to be employed with increasing frequency in health services. Some may believe it will produce the *Star Trek*-like medical suite in which the doctor runs the machine across the length of their patient's body and it spits out an answer. AI has already been shown to be better at picking up abnormalities on X-rays than many radiologists. That gives the impression that it's superior to medical staff at making a diagnosis. But there is a big difference, as I hope I have demonstrated, between finding an abnormality on a scan and knowing what it means to the patient's symptoms. Certainly, if you feed enough good-quality information into any AI algorithm it might also learn the nuance of diagnosis over time, but the results it produces will always depend on the information it's given. That information will always have to come from someone and that person needs to be able to listen to the story and do justice to the subtleties it contains. AI might help pathologists and radiologists speed through scans and specimens, but it has little hope of replacing an experienced person as diagnostician. And it cannot do the vital laying on of hands. Michael Balint, renowned psychoanalyst, put it best when he said that the drug most used in a doctor's surgery is the doctor himself. Sometimes it is not the details of the medical complaint but the act of complaining that a patient needs. In this, a machine or an algorithm is a long way off replacing a person.

Doctors need to rethink their hyperspecialised silos. There is great value in a specialist because no doctor can know everything about everything. The more specialist a doctor is in your disease, the better care you will get for that disease. But there is a downside. Very specialist doctors often become deskilled in general medicine to the detriment of diagnosis. It is difficult to come up with alternative explanations for a patient's symptoms if the only ones you know are in your own small wheelhouse. To a hammer, everything looks like a nail. The specialist system makes it far too easy for a doctor to see a patient as a single body part. It also makes it easy to ignore all the other diagnoses our patients are accruing. Even when we do note the growing diagnosis list, we are disempowered from treating the patient as a whole. Specialists don't question the practices of other specialists.

We need to learn to value the role of the generalist doctor once again. They have been somewhat sidelined by the view that specialists are the real experts. But specialists don't always have the full panoramic view of the patient. The hospital generalist and the primary care doctor have that. They know their patients as a whole person. They are the one most able to stop one person getting several diagnoses for the same problem. They know which patient is unlikely to gain by a referral to their umpteenth specialist. They cut back on polypharmacy, where a patient takes multiple medications for numerous medical conditions, sometimes with one medication serving no other purpose than to alleviate the side effects of another. We need to return to a system in which we have more overseeing generalist doctors who are in a position to notice when too much diagnosis and too much medicine has started to make a person worse instead of better.

At any point along Darcie's journey to ten different diagnoses, none of which made her any better, she could have benefitted from a doctor who had the power to stop the specialist-to-specialist referrals and consider whether a new strategy was needed. She would have been better served by an approach that de-escalated her worry about her symptoms and took attention away from her body. But it didn't happen because all her specialists were focused only on finding an explanation for one subset of her symptoms.

Specialists are stakeholders in a way that primary care and generalist doctors are not. Their careers live or die on how many diagnoses they make of their specialist condition and how many patients they attract into their service. When researchers or specialists or specialist committees decide that the public would benefit by a change of diagnostic criteria that improves recognition of milder and atypical cases of 'their' disease of interest, they are stakeholders here too. They stand to gain directly from more patients.

As I write, new criteria are being drawn up to make autism a more available diagnosis for women. This work is being done *before* we have established whether diagnosing women with autism makes their lives better or worse. Those drawing up the new diagnostic criteria are the same people who publish research on autism, who run services for people with autism, whose careers, reputations and incomes will grow the more people with autism they find. That is not a good model for reframing diagnosis and it is a system in need of reform. Every new committee should have to put their proposed changes to the overdiagnosis test, meaning they have to have a good sense of how many people might be affected by any change in disease definition and a measure of the benefit-to-harm ratio their new population of patients will face before the change is made. The evaluation of the efficacy of treatment strategies should happen before the redefinition expands the patient population. And, crucially, non-stakeholders – generalists, primary care doctors – should be involved in all these decisions. New diagnostic criteria need to be measured more by their ability to make quality of life better – not by how many patients they can find. If the new criteria for a female phenotype of autism are so inclusive as to identify millions of new people with autism worldwide, that will be considered to indicate success. But it shouldn't be. Success should be measured in real quality of life improvements beyond the short-term payoff of a label.

And as patients, we need to learn what good medicine looks like. It might be useful to start by dispelling our belief that technical medicine is superior to clinical medicine. Test results are not as clear cut as many people believe. A diagnosis comes from a story and an examination and a collaboration between two people: the patient and doctor. Even I had to be reminded of that lesson as I wrote this book. I had no idea that the exact same gene variant could convey a very high risk of cancer to one person but a much lower risk to others. Clinical diagnosis has always been central to neurology, but I somehow felt inclined to believe that perhaps other specialities were different, especially those subject to particularly impressive technical advances like the field of genetics has been. But no, medicine remains an art for all of us. I can get the most out of my doctor by trusting them to know what tests will benefit me and which won't.

It is in ordering tests that the *folie à deux* is seen very plainly, because medical professionals will provide the tests their patients want even if they know they're unnecessary and might potentially cause confusion. We even order tests for our patients that we would not agree to if we were the patient. One study showed that nearly 73% of physicians considered unnecessary tests and procedures to be a serious problem in medicine.[16] Even so, nearly 50% of those same physicians admitted to ordering at least one unnecessary test every week. Fear of being sued and patient insistence were the main reasons that doctors gave for sending patients for tests against their best instincts. Doctors are rarely sued for overdiagnosis so it can be a tempting route for us. Doctors also do tests if they don't have enough time. It can be much easier and quicker to book that unnecessary brain scan than to spend the time explaining why it's unnecessary. Doing tests is perceived as good medicine by some patients. Good doctors know that isn't the case but, under pressure, we sometimes agree to them anyway.

In 2002, Italian cardiologist Alberto Dolara suggested that 'slow medicine' was the solution to overdiagnosis.[17] Time was needed to allow a holistic evaluation of a person's needs, to reduce anxiety while waiting for non-urgent diagnostic and therapeutic procedures, to prevent premature dismissals from the hospital and to offer adequate emotional support. Research has shown that more time and attention from medical professionals for their patients leads to better diagnostics and a higher appreciation by patients. We need more doctors, nurses, psychologists, occupational therapists and physiotherapists, not more machines.

I am aware, of course, that any argument that suggests we should do fewer tests and less screening might scare people. It feels like rationing. It feels like going back in time. It isn't. A US study published in the *New England Journal of Medicine* in 2017 provides food for thought.[18] It compared low- and high-income countries in terms of life expectancy and confirmed, as one might expect, that a higher income was associated with greater longevity. However, it did not attribute better health to better medical care. Instead, it raised concerns that wealthier people receive too much care. The study found that the financially better off had a much higher rate of cancer diagnosis, probably because they were undergoing a great many more tests than poorer communities. However, the death rates from cancer were similar in low-income and high-income communities. This suggests that a significant proportion of those cancers diagnosed in the wealthy were overdiagnosed. The study estimated that for every life saved by cancer screening in a wealthy community, up to ten people underwent cancer treatment they didn't really need. The authors concluded, 'Some of the resistance to moving toward a more sustainable (and affordable) healthcare system comes from people who fear they will be forced to give something up. Our findings offer the possibility that what may be given up is *unnecessary* care.'

Fewer tests do not necessarily mean less care. It could mean replacing low-value care with something invaluable – time spent with healthcare professionals. Addressing overdiagnosis and overmedicalisation would allow this by freeing up considerable funds for health services. One study has suggested that

30% of money spent on healthcare in the US is of no benefit.[19] In 2013, a conservative analysis estimated that a minimum of $270 billion was spent on overuse, even though a significant part of the population still lacks access to healthcare services.[20] In the UK, 20% of clinical work is thought to have no effect on outcome.[21] In Australia, overmedicalisation has been identified as a bigger cause of health cost increases than population growth or ageing.[22]

The history of medicine is dotted with assumptions long held and trusted that later proved to be wrong. For a hundred years, there was a conviction that an enlarged thymus gland caused sudden infant death syndrome (SIDS). The practice of irradiating the apparently enlarged thymic glands of children to prevent SIDS didn't end until the 1940s, after it was discovered that those treated had a higher rate of breast and thyroid cancer. Frontal lobotomies used to be an entirely acceptable practice. Such was the excitement around this surgical remedy that the surgeon who pioneered the procedure even won the Nobel prize for it. It was used to treat all manner of mental health and behavioural problems, only falling out of favour in the 1950s. Stress was a long-running explanation for stomach ulcers until, in the early 1980s, the bacteria *Helicobacter pylori* was uncovered as the actual cause for most cases of chronic gastritis. Until the late 1990s, bed rest was assumed to be the best treatment for low back pain. For the near entirety of my junior doctor years, I told everyone with low back pain to stay in bed. Now we recommend the exact opposite, that people stay active and do gentle exercise and stretches.

Looking back, it is easy to be horrified that people with problems like depression had major brain surgery by way of 'treatment'. But, back then, that was what cutting-edge medicine looked like. It makes one wonder which diagnoses and treatments we believe in now will look equally shocking or inappropriate at some point in the future. It is impossible to know but one has to ask: will overenthusiastic, untested screening programmes, risk-reducing surgeries, scans for everyone, widespread genetic testing be among them? Will we laugh at how we fell in love with technology, using it on every person possible before we even really knew how to use it? Will we be amazed at how much energy we put into finding diseases that we knew we couldn't treat? Will we be disappointed by how often we told children they had brains that were neurodevelopmentally different?

We will surely find it odd how we repeated our mistakes. Nearly every week, I see a new press announcement for a test that will diagnose disease earlier than ever before. A blood test that detects multiple types of cancer, even though we still haven't done the work that proves our existing cancer screening programmes are working. A blood test that can predict the development of Alzheimer's disease ten years before the symptoms start, and another to detect pre-symptomatic Parkinson's disease, even though there is no disease-modifying therapy that will alter the downhill trajectory of either of these two diseases. Early diagnosis will certainly benefit science, researchers and future patients, but it will not necessarily help the people being diagnosed *now*. That is something that the press announcements and the calls for patients to come forward for testing rarely make clear. Supporters of these innovations (researchers, stakeholders) say that people are empowered by early diagnosis and people want to know. But I'm not sure that patients always realise that the treatments that are in development could be decades away from being completed. I'm quite sure that the medical community hasn't learned the lesson of the Huntington's disease community, who often prefer not to know of their impending disease.

If our aim is to create a healthier population, there may be better ways to do it than early diagnosis and screening. Antibiotics, vaccines, antimalarials, insulin and blood transfusions have saved billions of lives. But even more lives have been saved by good sanitation, better nutrition and advances in agricultural technology. Healthcare resources would be better spent on campaigns to help people live healthier lives. In the introduction to this book, I referred to a change in the definition of pre-diabetes that could potentially reclassify nearly 50% of the adult Chinese population as pre-diabetic.[23] This was a six-fold increase when compared with figures using older diagnostic criteria.[24] Once diagnosed, all those people could be subjected to regular monitoring and advice on weight management and diet. Wouldn't a better strategy be a public health campaign aimed at the entire population, rather than the expensive and anxiety-provoking medicalisation of 50%? Obesity, poor diet, inactivity and smoking all contribute to a person's risk of developing diabetes and all are available to improvement through social change.

I'm a doctor but I'm also a patient. I have all this medical knowledge, but I still struggle with knowing when I should agree to tests and how much screening I should accept for myself. I make those decisions as everyone should – after careful consideration and in partnership with my doctors. The key to getting good care is to know what good care looks like. It isn't more tests but rather appropriate testing carried out in the knowledge that what stands to be gained outstrips the risk of incidental findings. It is testing that takes the clinical context into account. It isn't a doctor who always agrees with you and does every test you want them to, but a doctor who listens and responds. A diagnosis should never be made just for the diagnosis' sake. It should come with some palpable improvement that broadens life's possibilities. We should seek wellness not through a diagnosis that reinforces what we *cannot do*, but through hobbies, interests, passions and social networks that remind us what we *can do*.

Othering places people into hierarchical groups – them and us. Part of the drive for improved diagnosis is to address the need for better recognition of suffering and to reduce stigma. But the medicalisation of human experience and over-inclusive diagnoses don't reduce stigma. They promote intolerance by othering; by dividing the world into neurodivergents and neurotypicals; by turning anything but optimal ageing into a disease; by creating a potential genetic underclass of people who are just waiting to be sick.

I started this book with the story of Huntington's disease even though I knew most people's lives would never be touched by it. HD may be uncommon but the experiences of people at risk of HD, those who could avail of an advanced diagnosis if they wanted it, resonated very strongly with me. It spoke to the invaluable nature of hope and how it sustains us. The American philosopher Ralph Waldo Emerson is credited with saying, 'It's not the destination, it's the journey.' It must be very hard to set out or continue on life's big journey if your aspirations have been narrowed by a diagnosis that is uncertain or offers very little. Let's leave diagnosis for those who are unequivocally sick and find a way to be more tolerant of difference and imperfections that still allows people to live an unencumbered life.

Acknowledgements

In writing this book I spoke to dozens of incredible people who gave me a great deal of their time. Thank you for so many very moving stories and for all your generosity. I wish I could have written every single story I was told, but it just wasn't possible. However, even those stories that were not included contributed a great deal to this book – I learned something from every person I spoke to and am endlessly grateful to each and every one. Particular thanks to 'Stephanie', who continues to send me samples of her poems and paintings, which always seem to arrive just when I need them most.

Numerous professionals – doctors, nurses, psychologists, scientists and others – helped me in my research, both by answering my many questions and directing me to educational resources. Some asked to remain anonymous and others I have not named because their contribution was specific to their own area of work and I did not want to presume that they would agree with and wish to be associated with every single related discussion. In looking for expert opinions I was met by incredibly clever people but, of course, any mistakes and all conclusions are ultimately my own.

Thank you to those charities that helped by introducing me to people who would like their stories told. They include Lyme Disease Action (who I recommend as a great resource and fount of common sense for those struggling with a diagnosis of

Lyme disease), SWAN UK, the Huntington's Disease Association, Lyme Research UK, Ovarian Cancer Action and Genetic Alliance UK.

Thank you to Wendy and Todd Murray and Isabelle Zeidner – for talking to me about their inspirational mother and grandmother, Polly Murray. Thank you also to Melissa Graffe at the Yale University Library for facilitating my visit and to Isabelle for connecting us.

This book started in conversation with my brilliant editor at Hodder Press, Kirty Topiwala, who is both very intelligent and insightful. I owe her a great deal. When the work was difficult, she did an excellent job of seeing to the heart of the problem and solving it with one sentence. Indeed everybody at Hodder Press has been wonderful. Thanks in particular to Anna Baty, Liz Marvin, Lucy Buxton and Ian Allen for their forensic attention to detail. Thanks to Louise Court, Liv French, Melissa Grierson, Alice Morley, Rebecca Folland and Melis Dagoglu for all the effort they put into making sure that this book would be read by someone, somewhere. Thanks also to Steve Leard for his cover design.

Similarly I have been very fortunate to work with Thesis in the US. I really appreciated the whole team's clear enthusiasm for the project which was apparent from our very first meeting. Thank you to publisher Adrian Zackheim for that. And more thanks still to my editor Bria Sandford for her ideas and suggestions that felt instantly right every time and which certainly strengthened the final product. Thank you also to editor Megan Wenerstrom, to publicity team Amanda Lang, Lauren Ball and Taylor Williams, and to Brian Lemus, for his cover design.

As always I am indebted to my agent Kirsty McLachlan from Morgan Green Creatives for giving my writing career that vital start and for continuing to support me.

Finally, an apology. Writing this book was often challenging so I apologise to those people in my life who had to bear witness to that and I apologise for disappearing into hermitage during the researching and writing process. And a reminder for myself. Sometimes, a bit like after childbirth, a book finally realised has a way of wiping away any memory of how painful the delivery was. But writing books is hard so when people ask you if writing while also working full time as a doctor in the NHS is difficult, please say – it is hell at times. And, with that in mind, going forward, only take on projects that, like the subject of this book, really matter.

Notes

Introduction

1 Caroline Williams, 'ADHD: What's behind the recent explosion in diagnoses?' *New Scientist*, 2 May 2023, https://www.newscientist.com/article/mg25834372-000-adhd-whats-behind-the-recent-explosion-in-diagnoses/

2 'Autism Prevalence Rises Again, Study Finds', *New York Times*, https://www.nytimes.com/2023/03/23/health/autism-children-diagnosis.html

3 'PTSD Has Surged Among College Students', https://www.nytimes.com/2024/05/30/health/ptsd-diagnoses-rising-college-students.html

4 Deidre McPhillips, 'More than 1 in 6 adults have depression as rates rise to record levels in the US, survey finds', CNN, 17 May 2023, https://edition.cnn.com/2023/05/17/health/depression-rates-gallup/index.html

5 Cindy Gordon, 'Massive health wake up call: depression and anxiety rates have increased by 25% in the past year', *Forbes*, 12 February 2023, https://www.forbes.com/sites/cindygordon/2023/02/12/massive-health-wake-up-call-depression-and-anxiety-rates-have-increased-by-25-in-the-past-year/

6 'Asthma trends and burden', lung.org, https://www.lung.org/research/trends-in-lung-disease/asthma-trends-brief/trends-and-burden

7 Steven Ross Johnson, 'New cancer cases projected to top 2 million, hit record high in 2024', *US News*, 17 January 2024, https://www.usnews.com/news/health-news/articles/2024-01-17/new-cancer-cases-projected-to-hit-record-high-in-2024

8 'Dementia diagnoses in England at record high', NHS England, 22 July 2024, https://www.england.nhs.uk/2024/07/dementia-diagnoses-in-england-at-record-high
9 https://diabetesatlas.org/
10 'Cancer screening', Nuffield Trust, https://www.nuffieldtrust.org.uk/resource/breast-and-cervical-cancer-screening
11 I.B. Richman et al, 'Estimating Breast Cancer Overdiagnosis After Screening Mammography Among Older Women in the United States', *Annals of Internal Medicine,* 176 (9) (2023)
12 Mengmeng Li et al, 'The economic cost of thyroid cancer in France and the corresponding share associated with treatment of overdiagnosed cases', *Value in Health,* 26 (8) (2023) pp.1175–1182
13 Jasmine Just, 'Overdiagnosis: when finding cancer can do more harm than good', Cancer Research UK, 6 March 2018, https://news.cancerresearchuk.org/2018/03/06/overdiagnosis-when-finding-cancer-can-do-more-harm-than-good/
14 John S. Yudkin et al, 'The epidemic of pre-diabetes: the medicine and the politics', *British Medical Journal,* 349 (2014)

1. Huntington's Disease

1. A. Maat-Kievit, 'Paradox of a better test for Huntington's disease', *Journal of Neurology, Neurosurgery and Psychiatry*, 69 (2000) pp.579–583
2. Seymour Kessler et al, 'Attitudes of persons at risk for Huntington disease toward predictive testing', *American Journal of Medical Genetics*, 26 (2) (1987) pp.259–70
3. 'Why adults at risk for Huntington's choose not to learn if they inherited deadly gene', Science Daily/Georgetown University Medical Center, 16 May 2019, https://www.sciencedaily.com/releases/2019/05/190516103715.htm
4. Karen E. Anderson, 'The choice not to undergo genetic testing for Huntington disease: Results from the PHAROS study', *Clinical Genetics*,96 (1) (2019) pp.28–34
5. Sheharyar S. Baig et al, '22 Years of predictive testing for Huntington's disease: the experience of the UK Huntington's Prediction Consortium', *European Journal of Human Genetics*, 24 (10) (2016) pp.1396–402

6. Giovanni Pezzulo et al, 'Symptom perception from a predictive processing perspective', *Clinical Psychology in Europe*, 1 (4) (2019) pp.1–14
7. Anne-Catherine Bachoud-Lévi et al, 'International guidelines for the treatment of Huntington's disease', *Frontiers in Neurology*, 10;710 (2019)
8. Maria U. Larsson et al, 'Depression and suicidal ideation after predictive testing for Huntington's disease: A two-year follow-up study', *Journal of Genetic Counseling*, 15, (5) (2006) pp.361–74
9. Robin McKie, 'Woman who inherited fatal illness to sue doctors in groundbreaking case', *Guardian,* 25 November 2018, https://www.theguardian.com/science/2018/nov/25/woman-inherited-fatal-illness-sue-doctors-groundbreaking-case-huntingtons
10. Institute of Medicine (US) Committee on Assessing Genetic Risks, Andrews L.B., Fullarton J.E., Holtzman N.A., et al, editors, 'Assessing genetic risks: implications for health and social policy', National Academies Press (Washington DC), 1994
11. Harry Fraser et al, 'Genetic discrimination by insurance companies in Aotearoa New Zealand: experiences and views of health professionals', *New Zealand Medical Journal*, 136(1574) (2023) pp.32–52

2. Lyme Disease and Long Covid

1. *The Widening Circle: A Lyme Disease Pioneer Tells Her Story* (St Martin's Press, 1996)
2. Ibid.
3. 'Lyme disease', National Institute for Clinical Care Excellence, https://cks.nice.org.uk/topics/lyme-disease/
4. 'Lyme disease surveillance and data', CDC, 15 May 2024, https://www.cdc.gov/lyme/data-research/facts-stats/index.html
5. A. Tonks, 'Lyme wars' *British Medical Journal,* (2007) 335:910
6. 'Treatment and intervention for Lyme disease', CDC, 16 August 2024, https://www.cdc.gov/lyme/treatment/index.html
7. 'Lyme disease guidelines', National Institute for Clinical Care Excellence, 11 April 2018, https://www.nice.org.uk/guidance/ng95
8. Andrew Moore et al, 'Current guidelines, common clinical pitfalls, and future directions for laboratory diagnosis of Lyme

disease', United States, *Emerging Infectious Diseases*, 22 (7) (2016) pp.1169–1177

9. S. O'Connell, 'Lyme disease in the United Kingdom', *British Medical Journal*, 310 (6975) (1995) pp.303–8
10. https://www.ca4.uscourts.gov/Opinions/Unpublished/151420.U.pdf
11. David Whelan, 'Lyme Inc.', 16 July 2012, https://www.forbes.com/forbes/2007/0312/096.html
12. Takaaki Kobayashi et al, 'Misdiagnosis of Lyme disease with unnecessary antimicrobial treatment characterizes patients referred to an academic infectious diseases clinic', Open Forum Infectious Diseases, 6 (7) (2019)
13. Rakel Kling et al, 'Diagnostic testing for Lyme disease: Beware of false positives', *British Columbia Medical Journal*, 57 (9) (2015), pp.396-99
14. 'Lyme disease surveillance and data', CDC, 15 May 2024, https://www.cdc.gov/lyme/data-research/facts-stats/index.html
15. 'About tick bite-associated illness in Australia', Australian Government Department of Health, https://www.health.gov.au/our-work/dscatt/about
16. 'Statistics', Lyme Disease Association of Australia, https://lymedisease.org.au/lyme-in-australia/statistics/
17. 'What is "chronic Lyme disease?"', National Institute of Allergy and Infectious Diseases, https://www.niaid.nih.gov/diseases-conditions/chronic-lyme-disease
18. 'A critical appraisal of "chronic Lyme disease"'; *New England Journal of Medicine*, 357 (2007) pp.1422–1430
19. Ed Yong, 'Covid-19 can last for several months', *Atlantic*, 4 June 2020, https://www.theatlantic.com/health/archive/2020/06/covid-19-coronavirus-longterm-symptoms-months/612679/
20. Elisabeth Mahase, 'Covid-19: What do we know about "long covid"?', *British Medical Journal*, 370 (2020)
21. Jeremy Devine, 'The dubious origins of long Covid', *Wall Street Journal*, 22 March 2021, https://www.wsj.com/articles/the-dubious-origins-of-long-covid-11616452583
22. https://www.wearebodypolitic.com/
23. Elisa Perego et al, 'Why we need to keep using the patient made term "long Covid"', the BMJ Opinion, *British Medical Journal*,

1 October 2020, https://blogs.bmj.com/bmj/2020/10/01/why-we-need-to-keep-using-the-patient-made-term-long-covid/
24. Felicity Callard and Elisa Perego, 'How and why patients made Long Covid', *Social Science and Medicine,* 268 (2021)
25. A.V. Raveendran et al, 'Long COVID: An overview', *Diabetology & Metabolic Syndrome,* 15 (3) (2021) pp.869–875
26. 'The dubious origins of long Covid', *Wall Street Journal*
27. César Fernández-de-las-Peñas et al, 'Post-COVID-19 symptoms 2 years after SARS-CoV-2 infection among hospitalized vs nonhospitalized patients', JAMA Network Open, 5(11) (2022) e2242106
28. M. Heightman et al, 'Post-COVID-19 assessment in a specialist clinical service: a 12-month, single-centre, prospective study in 1325 individuals', *BMJ Open Respiratory Research* 8 (2021) e001041
29. Ellen J. Thompson et al, 'Risk factors for long COVID: analyses of 10 longitudinal studies and electronic health records in the UK', *Nature Communications,* 13 (3529) (2022)
30. Harry Crook et al, 'Long Covid – mechanisms, risk factors, and management', *British Medical Journal,* 374 (2021)
31. Jennifer Senior, 'What Not to Ask Me About My Long Covid', *Atlantic,* 15 February 2023, https://www.theatlantic.com/ideas/archive/2023/02/long-covid-symptoms-chronic-illness-disability/673057/
32. 'Post COVID-19 condition (Long COVID)', WHO, 7 December 2022, https://www.who.int/europe/news-room/fact-sheets/item/post-covid-19-condition
33. 'Prevalence of ongoing symptoms following coronavirus (COVID-19) infection in the UK', Office for National Statistics, 30 March 2023, https://www.ons.gov.uk/peoplepopulationandcommunity/healthandsocialcare/conditionsanddiseases/bulletins/prevalenceofongoingsymptomsfollowingcoronaviruscovid19infectionintheuk/30march2023
34. Mary Kekatos, 'About 18 million US adults have had long COVID: CDC', ABC News, 26 September 2023, https://abcnews.go.com/Health/18-million-us-adults-long-covid-cdc/story?id=103464362
35. C. Lemogne et al, 'Why the hypothesis of psychological mechanisms in long COVID is worth considering', *Journal of Psychosomatic Research,* 165: 111135 (2023)
36. Ari R. Joffe and April Elliott, 'Long COVID as a functional somatic symptom disorder caused by abnormally precise prior

expectations during Bayesian perceptual processing: A new hypothesis and implications for pandemic response', *SAGE Open Medicine,* 11 (2023)

37. Michael Fleischer,'Post-COVID-19 syndrome is rarely associated with damage of the nervous system: findings from a prospective observational cohort study in 171 patients', *Neurology and Therapy,* 11 (2022) pp.1637–1657
38. Sara Gorman and Jack Gorman, 'The role of psychological distress in long Covid', *Psychology Today,* 4 October 2022, https://www.psychologytoday.com/gb/blog/denying-the-grave/202210/the-role-psychological-distress-in-long-covid
39. Matthjew S. Durstenfeld et al, 'Long COVID symptoms in an online cohort study', *Open Forum Infectious Diseases,* 10 (2) (2023)
40. Siwen Wang et al, 'Associations of depression, anxiety, worry, perceived stress, and loneliness prior to infection with risk of post-COVID-19 conditions', *JAMA Psychiatry,* 79(11) (2022), pp.1081–1091
41. Vasiliki Tsampasian et al, 'Risk factors associated with post-COVID-19 condition: a systematic review and meta-analysis', *JAMA Internal Medicine,* 183(6) (2023) pp.566–580
42. Elaine Hill et al, 'Risk factors associated with post-acute sequelae of SARS-CoV-2 in an EHR cohort: A national COVID cohort collaborative (N3C) analysis as part of the NIH RECOVER program', the RECOVER Consortium, medRxiv prepreint (2022)
43. Elizabeth T. Jacobs et al, 'Pre-existing conditions associated with post-acute sequelae of Covid-19', *Journal of Autoimmunity,* 135 (2023)
44. Joel Selvakumar et al, 'Prevalence and characteristics associated with post-Covid-19 condition among nonhospitalized adolescents and young adults', JAMA Network Open, 6 (3) (2023)
45. Kelsey McOwat, et al, 'The CLoCk study: A retrospective exploration of loneliness in children and young people during the COVID-19 pandemic, in England,' *PLoS One.* 21; 18 (11) (2023)
46. Petra Engelmann et al, 'Risk factors for worsening of somatic symptom burden in a prospective cohort during the COVID-19 pandemic', *Frontiers in Psychology,* 13 (2022)
47. Mark Shevlin et al, 'Covid-19-related anxiety predicts somatic symptoms in the UK population', *British Journal of Health Psychology,* 25 (4) (2020) pp.875–882

48. Liron Rozenkrantz et al, 'How beliefs about coronavirus disease (COVID) influence COVID-like symptoms? A longitudinal study' *Health Psychology*, 41 (8) (2022) pp.519–526
49. Justina Motiejunaite et al, 'Hyperventilation: A possible explanation for long-lasting exercise intolerance in mild Covid-19 survivors?', *Frontiers in Physiology*, 11: 614590 (2021)
50. Michael C. Sneller et al, 'A longitudinal study of COVID-19 sequelae and immunity: baseline findings', *Annals of Internal Medicine*, 175 (7) (2022) pp.969–979
51. 'Risk factors for worsening of somatic symptom burden in a prospective cohort during the COVID-19 pandemic', *Frontiers in Psychology*
52. T. Fox et al, 'What is the evidence that "microclots" cause the post-COVID-19 syndrome, and is removal using plasmapheresis justified?', Cochrane, 26 July 2023, https://www.cochrane.org/CD015775/INFECTN_what-evidence-microclots-cause-post-covid-19-syndrome-and-removal-using-plasmapheresis-justified
53. Klaus J, Wirth and Carmen Scheibenbogen, 'Dyspnea in post-COVID syndrome following mild acute COVID-19 infections: potential causes and consequences for a therapeutic approach', *Medicina*, 58 (3) (2022) p.419
54. 'A longitudinal study of COVID-19 sequelae and immunity: baseline findings', *Annals of Internal Medicine*, 2022
55. Mattieu Gasnier et al, 'Comorbidity of long COVID and psychiatric disorders after a hospitalisation for COVID-19: a cross-sectional study', *Journal of Neurology, Neurosurgery & Psychiatry* 93 (2022) pp.1091–1098
56. S. A. Behnood et al, 'Persistent symptoms following SARS-CoV-2 infection amongst children and young people: A meta-analysis of controlled and uncontrolled studies', *Journal of Infection*, 84 (2) (2022), pp.158–170
57. Siweem Wang, 'Associations of depression, anxiety, worry, perceived stress, and loneliness prior to infection with risk of post-COVID-19 conditions', *JAMA Psychiatry*, 79 (11) (2022) pp.1081–1091
58. Grace Huckins, 'Is Long COVID Linked to Mental Illness?', *Slate*, 26 June 2023, https://slate.com/technology/2023/06/mental-illness-long-covid-body-mind.html

3. Autism

1. 'Data and statistics on autism spectrum disorder', CDC, 16 May 2024, https://www.cdc.gov/autism/data-research/index.html
2. Adam Kula, 'Idea that 5% of all Northern Ireland's children are autistic is "a fantasy" claims international expert', Newsletter.co.uk, 12 June 2023, https://www.newsletter.co.uk/education/idea-that-5-of-all-northern-irelands-children-are-autistic-is-a-fantasy-claims-international-expert-4178467
3. John Mac Ghlionn, 'Doctor who helped broaden autism spectrum "very sorry" for over-diagnosis', *New York Post*, 24 April 2023, https://nypost.com/2023/04/24/doctor-who-broadened-autism-spectrum-sorry-for-over-diagnosis/
4. Peter Stanford, 'Simon Baron-Cohen: "The treatment of autistic people is a scandal on the scale of infected blood"', *Telegraph*, 15 June 2024, https://www.telegraph.co.uk/news/2024/06/15/simon-baron-cohen-interview-autism-scandal-infected-blood/
5. The DSM 3 and 4 each have two editions, the original and a revised edition, taking it to seven editions in total
6. Robyn L. Young and Melissa L. Rodi, 'Redefining Autism Spectrum Disorder Using DSM-5: The Implications of the Proposed DSM-5 Criteria for Autism Spectrum Disorders', *Journal of Autism and Developmental Disorders* 44 (2014), pp. 758–765
7. 'The prevalence of autism (including Aspergers syndrome) in-school age children in Northern Ireland. Annual report 2023', Department of Health, 18 May 2023
8. https://www.goldenstepsaba.com/resources/what-country-has-the-highest-rate-of-autism
9. Patricia M. Dietz, 'National and state estimates of adults with autism spectrum disorder', *Journal of Autism and Developmental Disorders*, 50 (12) (2020) pp.4258–4266
10. Rachel Loomes et al, 'What is the male-to-female ratio in autism spectrum disorder? A systematic review and meta-analysis', *Journal of the American Academy of Child and Adolescent Psychiatry*, 56 (6) (2017) pp.466–474
11. 'A qualitative exploration of the female experience of autism spectrum disorder (ASD)', *Journal of Autism and Developmental Disorders*, vol. 49, iss. 6 (2019) pp.2389–2402

12. Victoria Milner et al, 'Evidence of increasing recorded diagnosis of autism spectrum disorders in Wales, UK: An e-cohort study', *Autism*, 26 (6) (2022) pp.1499–1508
13. 'Elon Musk reveals he has Asperger's on Saturday Night Live', BBC News, 9 May 2021, https://www.bbc.co.uk/news/world-us-canada-57045770
14. Chanel Georgina, 'Sir Anthony Hopkins says his autism diagnosis is nothing more than a "fancy label"', *Sunday Express*, 2 October 2022, https://www.express.co.uk/life-style/health/1676488/sir-Anthony-Hopkins-health-aspergers-autism-symptoms
15. John N. Constantino and Richard D. Todd, 'Autistic traits in the general population: a twin study', *Archives of General Psychiatry,* 60 (5) (2003) pp.524–530
16. Victoria Milner, 'A qualitative exploration of the female experience of autism spectrum disorder (ASD)', *Journal of Autism and Developmental Disorders*, 49 (6) (2019) pp.2389–2402
17. L. Kanner, 'Autistic disturbances of affective contact', *Nervous Child,* 2 (1943) pp.217–250
18. Kristen Bottema-Beutel et al, 'Adverse event reporting in intervention research for young autistic children', *Autism,* 25 (2) (2021) pp.322–335
19. Yu-Chi Chou et al, 'Comparisons of self-determination among students with autism, intellectual disability, and learning disabilities: A multivariate analysis', *Focus on Autism and Other Developmental Disabilities,* 32 (2) (2016) pp.124–132
20. Xueqin Qian et al, 'Differences in self-determination across disability categories: findings from national longitudinal transition study', *Journal of Disability Policy Studies,* 32 (4) (2012) pp.245–256
21. Rifat, Kerem Gurkan and Funda Kocak, 'Double punch to the better than nothing: physical activity participation of adolescents with autism spectrum disorder', *International Journal of Developmental Disabilities*, 69 (5) (2021) pp.697–709
22. Lee Jussim, 'Self-fulfilling prophecies: A theoretical and integrative review', *Psychological Review, 93* (4) (1986) pp.429–445
23. Eric Fombonne, 'Editorial: Is autism overdiagnosed?', *Journal of Child Psychology and Psychiatry*, 64 (5) (2023) pp.711–714

24. 'Anxiety and depression in children: Get the facts', CDC, https://www.cdc.gov/childrensmentalhealth/features/anxiety-depression-children.html
25. 'Rising ill-health and economic inactivity because of long-term sickness, UK: 2019 to 2023', Office for National Statistics, 26 July 2023, https://www.ons.gov.uk/employmentandlabourmarket/peoplenotinwork/economicinactivity/articles/risingillhealthandeconomicinactivitybecauseoflongtermsicknessuk/2019to2023
26. 'One in five children and young people had a probable mental disorder in 2023', NHS England, 21 November 2023, https://www.england.nhs.uk/2023/11/one-in-five-children-and-young-people-had-a-probable-mental-disorder-in-2023/
27. Jessica Morris, 'The rapidly growing waiting lists for autism and ADHD assessments', Nuffield Trust QualityWatch, https://www.nuffieldtrust.org.uk/news-item/the-rapidly-growing-waiting-lists-for-autism-and-adhd-assessments
28. 'Editorial: Is autism overdiagnosed?', *Journal of Child Psychology and Psychiatry*
29. 'Some NHS centres twice as likely to diagnose adults as autistic, study finds', University College London, 5 March 2024, https://www.ucl.ac.uk/news/headlines/2024/mar/some-nhs-centres-twice-likely-diagnose-adults-autistic-study-finds
30. 'Doctor who helped broaden autism spectrum "very sorry" for over-diagnosis', *New York Post*
31. Diego Aragon-Guevara, 'The reach and accuracy of information on autism on TikTok', *Journal of Autism and Development Disorders* (2023)
32. Ellie Iorizzo, 'Tallula Willis reveals autism diagnosis: "It's changed my life"', Yahoo, 18 March 2024; Kate Ng, '"It's fantastic": Melanie Sykes says she is "celebrating" her autism diagnosis', *Independent*, 6 December 2021
33. 'Does Bill Gates have autism?', Rainbow, 13 April 2024, https://www.rainbowtherapy.org/blogs-does-bill-gates-have-autism/
34. 'Does Tim Burton have autism or Asperger's?', Golden Steps ABA, 3 August 2023, https://www.goldenstepsaba.com/resources/does-tim-burton-have-autism
35. 'Famous Autistic People', On The Spectrum Foundation, https://www.onthespectrumfoundation.org/famous-people-with-asperger-s

36. Jack Shepherd, 'Robbie Williams "believes he has Asperger Syndrome"', *Independent*, 29 June 2018, https://www.independent.co.uk/arts-entertainment/music/news/robbie-williams-asperger-syndrome-radio-2-interview-autism-spectrum-a8422461.html
37. https://www.thetimes.com/uk/healthcare/article/rise-of-autism-makes-diagnosis-meaningless-6pgssfznt
38. 'Concerns about Spectrum 10K: Common Variant Genetics of Autism and Autistic traits', NHS Health Research Authority, 22 May 2022, https://www.hra.nhs.uk/about-us/governance/feedback-raising-concerns/spectrum-10k-update-19-may-2022/

4. The Cancer Gene

1. 'BRCA Exchange: Facts & stats', BRCA Exchange, https://brcaexchange.org/factsheet
2. Not all variants in the BRCA genes increase a person's risk of cancer. Many are benign and cause no health problems. A 'pathological' or 'high risk' variant is one that does confer an increased cancer risk.
3. https://www.cancer.gov/about-cancer/causes-prevention/genetics/brca-fact-sheet
4. 'Surgery to Reduce the Risk of Breast Cancer', National Cancer Institute, https://www.cancer.gov/types/breast/risk-reducing-surgery-fact-sheet
5. Sofía Luque Suárez et al, 'Immediate psychological implications of risk-reducing mastectomies in women with increased risk of breast cancer. A comparative study', *Clinical Breast Cancer*, S1526-8209 (2024)
6. Stephanie M. Wong et al, 'Counselling framework for germline *BRCA1/2* and *PALB2* carriers considering risk-reducing mastectomy', *Current Oncology, 31* (2024) pp.350–365
7. Amanda S. Nitschke et al, 'Non-cancer risks in people with *BRCA* mutations following risk-reducing bilateral salpingo-oophorectomy and the role of hormone replacement therapy: a review', *Cancers, 15 (3)* (2023) pp.711
8. Minal S. Kale and Deborah Korenstein, 'Overdiagnosis and overtreatment; how to deal with too much medicine', *Journal of Family Medicine and Primary Care*, 9(8) (2020)

9. 'Overdiagnosis in primary care: framing the problem and finding solutions', *British Medical Journal*, 362 (2018)
10. 'Thyroid cancer: zealous imaging has increased detection and treatment of low risk tumours', *British Medical Journal*, 347 (2013)
11. 'Prostate-specific antigen screening and 15-year prostate cancer mortality: a secondary analysis of the CAP randomized clinical trial', *JAMA,* 331(17) (2024), pp.1460–1470
12. Brigid Betz-Stablein and H. Peter Soyer, 'Overdiagnosis in Melanoma Screening: Is It a Real Problem?', *Dermatol Pract Concept.* 13(4) (2023); Katy J.L. Bell, 'Melanoma overdiagnosis: why it matters and what can be done about it.', *British Journal of Dermatology,* 187 (4) (2022), pp. 459–460.
13. Daniel Lindsay et al, 'Estimating the magnitude and healthcare costs of melanoma in situ and thin invasive melanoma overdiagnosis in Australia', *British Journal of Dermatology* (2024)
14. Ilana B. Richman et al, 'Estimating breast cancer overdiagnosis after screening mammography among older women in the United States', *Annals of Internal Medicine,* 176(9) (2023) pp.1172–1180
15. 'Screening for breast cancer with mammography', *Cochrane Database of Systematic Reviews*
16. Oleg Blyuss et al, 'A case-control study to evaluate the impact of the breast screening programme on breast cancer incidence in England', *Cancer Medicine,* 12 (2) (2023) pp.1878–1887
17. Michael Bretthauer et al, 'Estimated lifetime gained with cancer screening tests: a meta-analysis of randomized clinical trials', *JAMA Internal Medicine,* 183 (11) (2023) pp.1196–1203
18. Kelly Metcalfe et al, 'International trends in the uptake of cancer risk reduction strategies in women with a BRCA1 or BRCA2 mutation', *British Journal of Cancer* 121 (1) (2019) pp.15–21
19. Narendra Nath Basu et al, 'The Angelina Jolie effect: Contralateral risk-reducing mastectomy trends in patients at increased risk of breast cancer', *Scientific Reports,* 11 (1) (2021) p.2847
20. Federica Chiesa and Virgilio S. Sacchini, 'Risk-reducing mastectomy', *Minerva Obstetrics and Gynecology,* 68 (5) (2016) pp.544–7
21. J. Morgan et al, 'Psychosocial outcomes after varying risk management strategies in women at increased familial breast cancer risk: a mixed methods study of patient and partner outcomes', *Annals of The Royal College of Surgeons of England,* 106 (1) (2024) pp.78–91

22. Katja Keller et al, 'Patient-reported satisfaction after prophylactic operations of the breast', *Breast Care* (Basel), 14 (4) (2019) pp.217–223
23. 'International trends in the uptake of cancer risk reduction strategies in women with a BRCA1 or BRCA2 mutation', *British Journal of Cancer*
24. Caroline F. Wright et al, 'Assessing the pathogenicity, penetrance, and expressivity of putative disease-causing variants in a population setting', *American Journal of Human Genetics*, 104 (2019) pp.275–86
25. Lynn B. Jorde and Michael J. Bamshad, 'Genetic ancestry testing: what is it and why is it important?' *JAMA*, 323 (11) (2020) pp.1089–1090
26. Kirpal S. Panacer, 'Ethical issues associated with direct-to-consumer genetic testing', *Cureus*, 15 (6) (2023)
27. Rachel Horton et al, 'Direct-to-consumer genetic testing', *British Medical Journal*, 367 (2019)
28. Amit Sud, 'Realistic expectations are key to realising the benefits of polygenic scores', *British Medical Journal*, 380 (2023)
29. Kelly F.J. Stewart et al, 'Behavioural changes, sharing behaviour and psychological responses after receiving direct-to-consumer genetic test results: a systematic review and meta-analysis', *Journal of Community Genetics*, 9 (1) (2018) pp.1–18
30. Gareth J. Hollands et al, 'The impact of communicating genetic risks of disease on risk-reducing health behaviour: systematic review with meta-analysis', *British Medical Journal*, 352 (2016)
31. 'Hancock criticised over DNA test "over reaction"', BBC News, 21 March 2019, https://www.bbc.co.uk/news/health-47652060
32. 'Are genetic tests useful to predict cancer?', Hannah Devlin, 23 March 2019, https://www.theguardian.com/society/2019/mar/23/are-predictive-genetic-test-useful-to-predict-cancer-matt-hancock
33. Ephrem Tadele Sedeta et al, 'Breast cancer: Global patterns of incidence, mortality, and trends', *Journal of Clinical Oncology*, 41 (2023) pp.10528–10528
34. 'Watch and wait', Cancer Research UK, https://www.cancerresearchuk.org/about-cancer/treatment/watch-and-wait
35. Marc D. Ryser et al, 'Outcomes of Active Surveillance for Ductal Carcinoma in Situ: A Computational Risk Analysis', *Journal of the National Cancer Institute*, 108(5) (2015)

36. Kirsten McCaffery et al, 'How different terminology for ductal carcinoma in situ impacts women's concern and treatment preferences: a randomised comparison within a national community survey', *BMJ Open* 5(11) (2015);
37. Edward Davies, 'Overdiagnosis: what are we so afraid of?', *British Medical Journal*, 12 September 2013, https://blogs.bmj.com/bmj/2013/09/12/edward-davies-overdiagnosis-what-are-we-soafraid-of/

5. ADHD, Depression and Neurodiversity

1. 'General Prevalence of ADHD', chadd.org, https://chadd.org/about-adhd/general-prevalence/
2. Elie Abdelnour et al, 'ADHD diagnostic trends: increased recognition or overdiagnosis?', *Missouri Medicine*, 119 (5) (2022) pp.467–473
3. Douglas G.J. McKechnie et al, 'Attention-deficit hyperactivity disorder diagnoses and prescriptions in UK primary care, 2000–2018: population-based cohort study', *BJPsych Open*, 9 (4) (2023) e121
4. Luise Kazda et al, 'Attention deficit/hyperactivity disorder (ADHD) in children: more focus on care and support, less on diagnosis', *British Medical Journal*, 384 (2024) e073448
5. Mohammad Al-Wardat et al, 'Prevalence of attention- deficit hyperactivity disorder in children, adolescents and adults in the Middle East and North Africa region: a systematic review and meta-analysis', *British Medical Journal Open*, 14 (2024) e078849 https://bmjopen.bmj.com/content/bmjopen/14/1/e078849.full.pdf
6. https://www.washingtonpost.com/national/health-science/adhd-about-1-in-5-adults-may-have-a-disorder-usually-associated-with-grade-school/2013/12/13/34634f4a-5b7f-11e3-a49b-90a0e156254b_story.html
7. 'Attention deficit hyperactivity disorder: How common is it?', National Institute of Clinical Excellence, https://cks.nice.org.uk/topics/attention-deficit-hyperactivity-disorder/background-information/prevalence/
8. https://www.theguardian.com/society/2023/oct/29/adult-adhd-autism-assessment-nhs-screening-system-yorkshirepilot#:~:text=The%20ADHD%20Foundation%20has%20indicated,services%20have%20struggled%20to%20cope

9. Eleni Frisira et al, 'Systematic review and meta-analysis: relative age in attention-deficit/ hyperactivity disorder and autism spectrum disorder', *European Child and Adolescent Psychiatry,* (2024); Martin Whitely et al, 'Annual Research Review: Attention deficit hyperactivity disorder late birthdate effect common in both high and low prescribing international jurisdictions: a systematic review', *Journal of Child Psychology and Psychiatry,* 60(4) (2019), pp.380–391
10. Tarjei Widding-Havneraas et al, 'Geographical variation in ADHD: do diagnoses reflect symptom levels?', *European Child and Adolescent Psychiatry*, 32 (9) (2023) pp.1795–1803
11. 'State-based Prevalence of ADHD Diagnosis and Treatment 2016–2019', CDC, https://www.cdc.gov/adhd/data/state-based-prevalence-of-adhd-diagnosis-and-treatment-2016-2019.html
12. James J. McGough, 'Psychiatric comorbidity in adult attention deficit hyperactivity disorder: Findings from multiplex families', *American Journal of Psychiatry*, 162(9) (2005) pp.1621–7
13. ADDitude editors, 'What Is ADHD? Symptoms, Subtypes & Treatments', ADDitude, 26 September 2019, https://www.additudemag.com/what-is-adhd-symptoms-causes-treatments/
14. 'What is ADHD/ADD?', ADHD Ireland, https://adhdireland.ie/general-information/what-is-adhd/
15. Oliver Grimm et al, 'Genetics of ADHD: what should the clinician know?', *Current Psychiatry Reports*, 22 (4) (2020) p.18; Sami Timimi, 'Insane Medicine, Chapter 3: The Manufacture of ADHD (Part 2)', Mad in America, 16 November 2020, https://www.madinamerica.com/2020/11/insane-medicine-chapter-3-manufacture-adhd-part-2a/
16. Judy Singer interview played on *AntiSocial* with Adam Fleming, BBC Radio 4, 27 January 2023
17. 'Seven-fold increase in adult ADHD prescriptions over 10 years', BBC News, 28 August 2023, https://www.bbc.co.uk/news/uk-scotland-66135145
18. Ben Beaglehole, 'Despite a tenfold increase in ADHD prescriptions, too many New Zealanders are still going without', The Conversation, 2 May 2024, https://theconversation.com/despite-a-tenfold-increase-in-adhd-prescriptions-too-many-new-zealanders-are-still-going-without-229179

19. Shannon Brumbaugh et al, 'Trends in characteristics of the recipients of new prescription stimulants between years 2010 and 2020 in the United States: An observational cohort study', eClinicalMedicine 50 (2022) 101524
20. R. Thomas, 'Attention deficit/Hyperactivity disorder: are we helping or harming?' *BMJ* (2013)
21. K. Boesen et al, 'Extended-release methylphenidate for attention deficit hyperactivity disorder (ADHD) in adults', Cochrane, 24 February 2022, https://www.cochrane.org/CD012857/BEHAV_extended-release-methylphenidate-attention-deficit-hyperactivity-disorder-adhd-adults
22. Joanna Moncrieff et al, 'The serotonin theory of depression: a systematic umbrella review of the evidence', *Molecular Psychiatry*, 28 (2023) pp.3243–3256
23. Susan Mayor, 'Meta-analysis shows difference between antidepressants and placebo is only significant in severe depression', *British Medical Journal*, 336 (2008) p.466
24. 'Position statement on antidepressants and depression', Royal College of Psychiatrists, May 2019, https://www.rcpsych.ac.uk/docs/default-source/improving-care/better-mh-policy/position-statements/ps04_19---antidepressants-and-depression.pdf?sfvrsn=ddea9473_5
25. 'Depression: Learn More – How effective are antidepressants?', National Library of Medicine, https://www.ncbi.nlm.nih.gov/books/NBK361016
26. 'One in five children and young people had a probable mental disorder in 2023', NHS England, 21 November 2023, https://www.england.nhs.uk/2023/11/one-in-five-children-and-young-people-had-a-probable-mental-disorder-in-2023/
27. J. Dykxhoorn et al, 'Temporal patterns in the recorded annual incidence of common mental disorders over two decades in the United Kingdom: a primary care cohort study', *Psychological Medicine*. 54(4) (2024), pp. 663–674
28. Ágnes Zsila and Marc Eric S. Reyes, 'Pros & cons: impacts of social media on mental health', *BMC Psychology*, 11 (2023) p.201
29. Laura Marciano et al, 'Digital media use and adolescents' mental health during the Covid-19 pandemic: a systematic review and meta-analysis', *Frontiers in Public Health*, 9 (2022) 793868

30. Ruth Plackett et al, 'The longitudinal impact of social media use on UK adolescents' mental health: longitudinal observational study', *Journal of Medical Internet Research*, 25 (2023) e43213
31. Andree Hartanto et al, 'Does social media use increase depressive symptoms? A reverse causation perspective', *Frontiers in Psychiatry*, Sec. Public Mental Health, 12 (2021)
32. Hasan Beyari and Sen-Chi Yu, 'The relationship between social media and the increase in mental health problems', *International Journal of Environmental Research and Public Health*, 20 (3) (2023), p.2383
33. Sharon Neufeld, senior research fellow, Cambridge University, interviewed on *The Briefing Room*, BBC, 22 July 2024
34. 'The power threat meaning framework', The British Psychological Society, January 2018, https://cms.bps.org.uk/sites/default/files/2022-10/PTMF%20overview.pdf
35. Liesbet Van Bulck et al, 'Illness identity: A novel predictor for healthcare use in adults with congenital heart disease', *Journal of the American Heart Association*, 7 (11) (2018)
36. Veronica W. Wanyee and Dr Josephine Arasa, 'Literature review of the relationship between illness identity and recovery outcomes among adults with severe mental illness', *Modern Psychological Studies*, 25 (2) (2020), https://scholar.utc.edu/cgi/viewcontent.cgi?article=1513&context=mps
37. Paul Garner, 'Paul Garner on long haul Covid-19 – Don't try to dominate this virus, accommodate it', The BMJ Opinion,4 September 2020, https://blogs.bmj.com/bmj/2020/09/04/paul-garner-on-long-haul-covid-19-dont-try-and-dominate-this-virus-accommodate-it/
38. Paul Garner, 'Paul Garner: For 7 weeks I have been through a roller coaster of ill health, extreme emotions, and utter exhaustion', The BMJ Opinion, 5 May 2020, https://blogs.bmj.com/bmj/2020/05/05/paul-garner-people-who-have-a-more-protracted-illness-need-help-to-understand-and-cope-with-the-constantly-shifting-bizarre-symptoms/
39. 'Paul Garner on long haul Covid-19 – Don't try to dominate this virus, accommodate it', The BMJ Opinion
40. J. Biederman, 'Attention deficit/hyperactive disorder: a lifespan perspective', *Journal of Clinical Psychiatry*, 59 (supplement 7) (1998) pp.4–16

41. R. Gittelman et al, 'Hyperactive boys almost grown up. I. Psychiatric status'. *Archives of General Psychiatry*, 42 (10) (1985) pp.937–47
42. Mélodie Lemay-Gaulin, 'Efficacy and Perceptions of Academic Accommodations for University Students with ADHD', doctoral thesis, August 2022, https://papyrus.bib.umontreal.ca/xmlui/bitstream/handle/1866/27699/Lemay-Gaulin_Melodie_Essai.pdf
43. Benjamin J. Lovett and Jason M. Nelson, 'Educational accommodations for children and adolescents with attention-deficit/hyperactivity disorder', *Journal of the American Academy of Child & Adolescent Psychiatry*, 60, (4) (2021) pp.448–457
44. 'Academic testing accommodations for ADHD: Do they help?', Learn Disability Association of America, 21 (2) (2016) pp.67–78
45. Dorien Jansen et al, 'The implementation of extended examination duration for students with ADHD in higher education', *Journal of Attention Disorders*, 23 (14) (2018) pp.1746–1758
46. Kapil Sayal et al, 'Impact of early school-based screening and intervention programs for ADHD on children's outcomes and access to services: Follow-up of a school-based trial at age 10 years', *Archives of Pediatrics & Adolescent Medicine*, 164 (5) (2010) pp.462–9
47. Franco De Crescenzo et al, 'Pharmacological and non-pharmacological treatment of adults with ADHD: A meta-review', *Evidence Based Mental Health*, 20 (2017) pp.4–11
48. William E. Pelham et al, 'The effect of stimulant medication on the learning of academic curricula in children with ADHD: A randomized crossover study', *Journal of Consulting and Clinical Psychology*, 90 (5) (2022) pp.367–380
49. Janet Currie et al, 'Do stimulant medications improve educational and behavioral outcomes for children with ADHD?' *Journal of Health Economics*, 37 (2014) pp.58–69
50. Perry T., editor, Therapeutics Letter, Vancouver (BC): Therapeutics Initiative; 1994-, Letter 110, 'Stimulants for ADHD in children: Revisited', 2018 Feb.
51. Samuele Cortese, 'Evidence-based prescribing of medications for ADHD: Where are we in 2023?', *Expert Opinion on Pharmacotherapy*, 24 (4) (2023) pp.425–434
52. Owens, J., Jackson, H., 'Attention-deficit/hyperactivity disorder severity, diagnosis, and later academic achievement in a national sample', *Social Science Research*, 61 (2017) pp.251–265

6. Syndrome Without a Name

1. 100,000 Genomes Project, https://www.genomicsengland.co.uk/initiatives/100000-genomes-project
2. 'BRCA Exchange: Facts and Stats', BRCA Exchange, https://brcaexchange.org/factsheet
3. Incidental findings and borderline results are a very common outcome of tests. If I do a large range of blood tests, they almost never come back without one borderline result. It is standard practice for doctors to use their clinical judgement to decide which need to be discussed with the patient and which are so inconsequential that they do not need to be passed on.
4. 'Newborn Genomes Programme', Genomics England, https://www.genomicsengland.co.uk/initiatives/newborns
5. Guardian Study, https://guardian-study.org/
6. https://babyscreen.mcri.edu.au/about/
7. Nina B. Gold et al, 'Perspectives of rare disease experts on newborn genome sequencing', *JAMA Network Open*, 6 (5) (2023) e2312231
8. Suzannah Kinsella et al, 'A Public Dialogue to Inform the Use of Wider Genomic Testing When Used as Part of Newborn Screening to Identify Cystic Fibrosis', *International Journal of Neonatal Screening,* 8(2) (2022)
9. Emma Wilkinson, 'Newborn genome screening: a step too far?', *Pharmaceutical Journal*, 6 January 2023, https://pharmaceutical-journal.com/article/feature/newborn-genome-screening-a-step-too-far
10. 'Non-invasive prenatal testing (NIPT)', NHS Inform, https://www.nhsinform.scot/healthy-living/screening/pregnancy/non-invasive-prenatal-testing-nipt/#how-nipt-works
11. 'NIPT Test', Cleveland Clinic, https://my.clevelandclinic.org/health/diagnostics/21050-nipt-test
12. 'Non-invasive prenatal testing for Down's Syndrome is 99% accurate and is preferred by parents', Great Ormond Street Hospital for Children, 9 June 2015, https://www.gosh.nhs.uk/press-releases/non-invasive-prenatal-testing-downs-syndrome-99-accurate-and-preferred-parents-0/
13. Catherine Joynson, 'Our concerns about non-invasive prenatal testing (NIPT) in the private healthcare sector', Nuffield Council on Bioethics, 8 February 2019, https://www.nuffieldbioethics.org/blog/nipt-private

14. 'Non-invasive prenatal testing: ethical issues', Nuffield Council on Bioethics, March 2017, https://www.nuffieldbioethics.org/assets/pdfs/NIPT-ethical-issues-full-report.pdf
15. 'Our concerns about non-invasive prenatal testing (NIPT) in the private healthcare sector', Nuffield Council on Bioethics
16. 'Fact checking: Non-invasive prenatal testing (NIPT) for Down's syndrome', Down's Syndrome Association, https://www.downs-syndrome.org.uk/wp-content/uploads/2020/08/2020.FactChecker_NIPT.pdf
17. Ibid.; Van Der Miej et al, 'TRIDENT-2: National Implementation of Genome-wide Non-invasive Prenatal Testing as a First-Tier Screening Test in the Netherlands', *The American Journal of Human Genetics*, 105 (6), pp.1091–1101
18. Zainab Al-Ibraheemi et al, 'Changing face of invasive diagnostic testing in the era of cell-free DNA', *American Journal of Perinatology*, 34 (11) (2017) pp.1142–1147
19. EUROCAT Working Group, 'Survey of prenatal screening policies in Europe for structural malformations and chromosome anomalies, and their impact on detection and termination rates for neural tube defects and Down's syndrome', *BJOG*, 115 (6) (2008) pp.689–96
20. 'FactCheck: Are 90% of babies with Down syndrome in Britain aborted?', *The Journal*, 3 February 2018, https://www.thejournal.ie/factcheck-babies-abortion-3823611-Feb2018/
21. Sarina R. Chaiken et al, 'Association between rates of Down syndrome diagnosis in states with vs without 20-week abortion bans from 2011 to 2018', *JAMA Network Open*, 6 (3) (2023) e233684
22. Julian Quinones and Arijeta Lajka, '"What kind of society do you want to live in?": Inside the country where Down syndrome is disappearing', CBC News, 15 August 2017, https://www.cbsnews.com/news/down-syndrome-iceland/
23. 'ASA bans prenatal testing ads for the use of misleading statistics', Nuffield Council on Bioethics, 20 November 2019, https://www.nuffieldbioethics.org/news/asa-bans-prenatal-testing-ads-for-the-use-of-misleading-statistics
24. 'When they warn of rare disorders, these prenatal tests are usually wrong', https://www.nytimes.com/2022/01/01/upshot/pregnancy-birth-genetic-testing.html

Conclusion

1. Darcisio Hortelan Antonio and Claudia Saad Magalhaes, 'Survey on joint hypermobility in university students aged 18–25 years old', *Advances in Rheumatology* 58 (3) (2018)
2. https://www.ehlers-danlos.com/what-is-eds/#:~:text=Classical%20EDS%20(cEDS)%20and%20vascular,1%20in%201%20million%20people
3. Joanne C. Demmler et al, 'Diagnosed prevalence of Ehlers-Danlos syndrome and hypermobility spectrum disorder in Wales, UK: a national electronic cohort study and case control comparison', *BMJ Open* 9 (2019) e031365
4. R.A. Wedge et al (eds), National Academies of Sciences, Engineering, and Medicine; Health and Medicine Division; Board on Health Care Services; Committee on Selected Heritable Disorders of Connective Tissue and Disability, 'Ehlers-Danlos Syndromes and Hypermobility Spectrum Disorders', National Academies Press (US), (2022), https://www.ncbi.nlm.nih.gov/books/NBK584966/
5. Cheryl Iny Harris, 'Covid-19 increases the prevalence of postural orthostatic tachycardia syndrome: What nutrition and dietetics practitioners need to know', *Journal of the Academy of Nutrition and Dietetics*, 122 (9) (2022) pp.1600–1605
6. Lesley Kavi, 'Postural tachycardia syndrome and long COVID: an update', *British Journal of General Practice*, 72 (714) (2022) pp.8–9
7. 'Osteoarthritis: Key facts', World Health Organization, 14 July 2023, https://www.who.int/news-room/fact-sheets/detail/osteoarthritis
8. Yixiang He et al, 'Global burden of osteoarthritis in adults aged 30 to 44 years, 1990 to 2019: Results from the Global Burden of Disease Study 2019', *BMC Musculoskeletal Disorders* 25 (2024) p.303
9. 'Two fifths of people have chronic pain by their 40s, with consequences for later life', University College London, 2 November 2022, https://www.ucl.ac.uk/news/2022/nov/two-fifths-people-have-chronic-pain-their-40s-consequences-later-life
10. Lucy Norcliffe-Kaufmann, 'Fear conditioning as a pathogenic mechanism in the postural tachycardia syndrome', *Brain*, 145(11) (2022) pp.3763–3769
11. Ian Hacking, 'Making up people', *London Review of Books*, 28 (16) (2006), https://www.lrb.co.uk/the-paper/v28/n16/ian-hacking/making-up-people

12. Jenny L.L. Csecs et al, 'Joint hypermobility links neurodivergence to dysautonomia and pain', *Frontiers in Psychiatry*, 12, 786916 (2022)
13. David Harris, 'Mast cell activation is linked to a wide range of other conditions', EDS.Clinic, https://www.eds.clinic/articles/mast-cell-activation-is-linked-to-a-wide-range-of-other-conditions
14. Emily L. Casanova et al, 'The Relationship between autism and Ehlers-Danlos syndromes/hypermobility spectrum disorders', *Journal of Personalized Medicine*, 10 (4) (2020) pp.260
15. Robert T. Thibault, 'Treating ADHD with suggestion: Neurofeedback and placebo therapeutics', *Journal of Attention Disorders*, 22 (8) (2018) pp.707–711
16. 'Unnecessary tests and procedures in the health care system', the ABIM Foundation/ PerryUndem Research/Communication, 1 May 2014, https://www.choosingwisely.org/files/Final-Choosing-Wisely-Survey-Report.pdf
17. 'A Brief History of Slow Medicine', Slow Medicine, 26 May 2019, https://www.slowmedicine.com.br/the-slow-medicine-history-by-ladd-bauer/
18. H. Gilbert Welch and Elliott S. Fisher, 'Income and cancer overdiagnosis – when too much care is harmful', *New England Journal of Medicine*, 376 (2017) pp.2208–2209
19. William H. Shrank et al, 'Waste in the US health care system: estimated costs and potential for savings', *JAMA*, 322(15) (2019) pp.1501–1509
20. Shannon Brownlee et al, 'Evidence for overuse of medical services around the world', *Lancet*, 390(10090) (2017) pp.156–168
21. Hugh Alderwick, 'Is the NHS delivering too much of the wrong things?', The King's Fund, 12 August 2015, https://www.kingsfund.org.uk/insight-and-analysis/blogs/nhs-delivering-too-much-wrong-things
22. João Pedro Bandovas et al, 'Broadening risk factor or disease definition as a driver for overdiagnosis: A narrative review', *Journal of Internal Medicine*, 291(4) (2022) pp.426 –437
23. John Yudkin, 'The epidemic of pre-diabetes: the medicine and the politics', *British Medical Journal*, 349 (2014)
24. John S. Yudkin, '"Prediabetes": Are There Problems With This Label? Yes, the Label Creates Further Problems!', *Diabetes Care* 39(8) (2016), pp. 1468–1471

Index